Patrick Moore's
Practical Astronomy Series

Springer
London
Berlin
Heidelberg
New York
Hong Kong
Milan
Paris
Tokyo

Other titles in this series

Observing Variable Stars

Gerry A. Good

With 55 Figures

Springer

QB
835
.G58
2003
Apr 2004

Cover illustration: An artist's impression of an eclipsing variable star.

British Library Cataloguing in Publication Data
Good, Gerry A.
    Observing variable stars. – (Patrick Moore's practical
    astronomy series)
    1. Variable stars – Observers' manuals 2. Variable stars –
    Amateurs' manuals
    I. Title
    523.8′44
    ISBN 1-85233-498-3

Library of Congress Cataloging-in-Publication Data
Good, Gerry A., 1954–
    Observing variable stars / Gerry A. Good.
        p. cm. – (Patrick Moore's practical astronomy series,
        ISSN 1617-7185)
    Includes bibliographical references and index.
    ISBN 1-85233-498-3 (alk. paper)
    1. Variable stars – Observers' manuals. I. Title. II. Series.
    QB835.G58 2003
    523.8′44–dc21   2002190895

Patrick Moore's Practical Astronomy Series ISSN 1617-7185
ISBN 1-85233-498-3 Springer-Verlag London Berlin Heidelberg
a member of BertelsmannSpringer Science+Business Media GmbH
http://www.springer.co.uk

Typeset by EXPO Holdings, Malaysia
Printed and bound at the Cromwell Press, Trowbridge, Wiltshire
58/3830-543210   Printed on acid-free paper   SPIN 10836542

Acknowledgments

I began this book as a simple project to formalize much of the information and methods that I wanted at my fingertips when I observe variable stars. Soon thereafter, it was suggested that I may want to share this information. I could have written a book three times this size but then no one would publish it.

No book is ever written by one person. This one is no exception so I would like to acknowledge the many people from whom I've received assistance, encouragement, and advice: Dr. John Percy, University of Toronto, a strong advocate of professional–amateur collaboration within the astronomy community, for pointing me in the right direction regarding professional-amateur partnerships; Dr. Robert Stebbins, University of Calgary, for providing me with his excellent research studies regarding professional–amateur partnerships; Dr. Joe Patterson, Columbia University and guiding mentor for the Center for Backyard Astrophysics, for allowing me to use CBA material in this book; Dr. Taichi Kato, Kyoto University, Japan, for assistance and kind words during the WZ Sagittae campaign and for permission to use VSNET data and charts; Dr. Tim Brown, High Altitude Observatory/ National Center for Atmospheric Research for allowing me to use data and charts; Dr. Michel Breger for allowing me to use data and information from the Delta Scuti network; Dr. Douglas Hall for allowing me to use IAPPP data; Dr. Janet Mattei, American Association of Variable Star Observers, for allowing me to use AAVSO charts and data; Dr. Arne Hendon, United States Naval Observatory, Flagstaff Station, who has answered many questions over several years and for providing much "public domain" advice; Gary Poyner, astronomer from Birmingham, England, and well-known cataclysmic variable observer, for suggesting topics concerning cataclysmic variable observations; Emile Schweitzer, past president of the AFOEV for allowing me to use the organization's variable star data; Roger Pickard, President of the British Astronomical Association Variable Star Section for allowing me to use BAA VSS data; Kari Tikkannen, Finland, for allowing me to use his data and charts; Olga V. Durlevich, *GCVS* research team, Sternberg Astronomical Institute, Moscow, Russia, for permission to use *GCVS* data and quotes; John Watson, my publisher for allowing me to proceed with kind guidance and encouraging words; Dr. Thomas Williamson, good friend and observing partner, whose contagious enthusiasm for astronomy has forced me to believe that sleep is really a luxury; Lisa Wood, good friend and observing partner, who has read and re-read many draft manuscripts of this book, provided myriad suggestions, and allowed me to spend hundreds of hours observing with her and sharing her successes and challenges so that we can all learn something from her experiences; and my wife, Jillian, who has tolerated months of clunking, clanking and drive motor noise during many late-night observing sessions, along with weekends spent analyzing light curves, reading manuscripts and patiently listening to me talk about variable stars.

Thank you all.

Contents

Chapter 1

Observing Variable Stars

We shall not cease from exploration
And the end of all our exploring
Will be to arrive where we started
And know the place for the first time.

T.S. Eliot

Set free your imagination and consider one of the remarkable curiosities of nature.

A star located 600 light-years from Earth is about to do something cataclysmic. In truth, it did so 600 years ago. However, as if looking through a time machine, you can see it happen tonight. It took 600 years for the light record of this amazing event to reach us, even traveling at 300,000 kilometers per second. Upon close examination of this star you will discover that it is really two stars, a *binary system*, but to your naked eye it appears as a single star because of its immense distance. As a coincidence of nature, one of the stars of this binary system is much like our Sun, a relatively small yellow star. However, the second star of this system is a much smaller *white dwarf*, only about as big as the Earth and it is bright blue. A million Earth-sized planets would fit into our Sun so compared to its larger companion this planet-sized dwarf is as a flea to a dog.

Remarkably, the two stars orbit each other in just a few hours. They are very close to each other; the second reason that the two stars appear as one star. Another amazing fact is that although the smaller star is only the size of the Earth it is as heavy as our Sun. If you could bring a piece of this star that was the size of a sugar cube to Earth, it would weigh 16 tons (tonnes). Even more impressive, because gravity is so much stronger

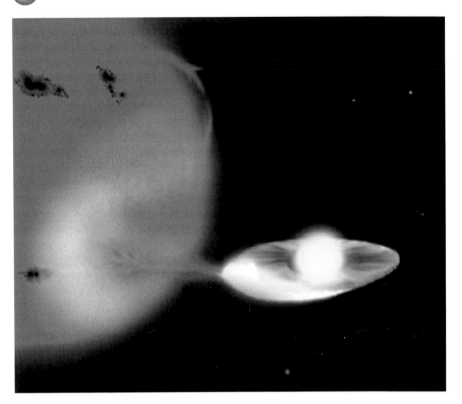

Figure 1.1. Artist's conception of a dwarf nova showing the tendril of hydrogen forming an accretion disk around the white dwarf star. Copyright: Gerry A. Good.

on the surface of the massive white dwarf, this sugar cube sized piece of star stuff weighs 470,000 times more at this star's surface than it would on the Earth's surface; an unbelievable 7.5 million tons!

As a result of the larger star being so close to the smaller star, as well as the massive nature of the dwarf star, gravity allows the smaller star to steal hydrogen from the larger one at an impressive rate. A tremendous amount of hydrogen is constantly being pulled from the larger star forming a gigantic tendril of streaming hot gas leading to a disk of stolen fuel that is now encircling the smaller star like a giant donut (Figure 1.1). Hydrogen is the fuel of stars and it is the primary source of energy that powers most of them. In this situation, the hydrogen will fuel a catastrophic explosion and the larcenous nature of the smaller star will be responsible for the eventual death of this binary system! But that will be in the far future.

Tonight, in a spectacular demonstration of these chaotic circumstances and every two weeks hence, after the disk of hydrogen gas has again grown to a sufficient size, the smaller star will proclaim its theft by producing an explosion within the disk that has the power of millions of nuclear bombs. As a result of the enormous explosion, this binary system becomes a hundred times brighter for a few hours.

Possibly a gift of nature, a reward for proper preparation, planning and patience, this dramatic sequence of events will be witnessed by a variable star observer somewhere on Earth. Maybe the observer will be you. To assist you in the preparation and planning for the observation of phenomena such as the one just described, this book will explain the basic nature of the various types of variable stars and suggest how best to observe them.

Why Do Amateur Astronomers Observe Variable Stars?

As with all amateur astronomers, you want to start seeing some results for your efforts as quickly as possible. You may have just spent a considerable amount of money for your new equipment and you want to start observing interesting things right now. Or you may have a pair of binoculars and wonder if there is anything happening up there, in the sky at night worthy of your attention. If you don't know where to look, the Universe can quickly seem like a rather boring place. Ironically, many would-be amateur astronomers give up very quickly because they cannot find anything interesting to view in a sky full of stars. A couple of nights under a star filled sky that doesn't seem to be *doing anything* and you'll believe that it's time to find a new hobby.

The Universe certainly *is* dynamic and ever-changing but most of the changes that take place occur on time-scales that exceed the life span of human beings and even the history of humankind. Watching a nebula expand or a star evolve will take millennia. On the other hand, trying to capture beautiful planetary nebulae, distant galaxies, planets within our own Solar System or the surface features of the Moon using astrophotography requires months, if not years, of effort to develop the necessary skills that will produce acceptable results. It is difficult to look out upon the Universe on any given night and see something happening. Without some help, you may give up, believing that there is nothing to see or that you lack the necessary time and skill to produce any results for your efforts.

This is where variable stars come to your rescue. As an amateur astronomer observing variable stars you will see immediate results from your labor. Cataclysmic variable stars *explode* within a few short hours, binary stars can eclipse one another several times a night and short-period variables change in brightness from evening to evening. These are events that you as an amateur astronomer can observe using binoculars or a telescope. More importantly, these are things that change quickly enough to give you some immediate satisfaction. Your binoculars will indeed show you something happening up there or the expense of a new telescope will seem justified after you actually witness the Universe change before your eyes. From night to night, there will be something new happening that you can see. In some cases, you will see change manifest itself over the span of several hours or, by creating light curves of these stars that vary, you will begin to see the nightly changes that you have been carefully recording for several weeks.

So, why do amateur astronomers observe variable stars? *Amateur astronomers observe variable stars because they can see changes occur every night!*

A Short Chronicle of Variable Star Observation

It is said that astronomy is one of two physical sciences in which the solitary amateur can make serious contributions, the other science being paleontology. For example, it should be readily apparent that an amateur nuclear physicist or an amateur genetic biologist will have great difficulties in contributing to their respective pastime regardless of their talent. This is not so for the amateur astronomer. Every night under the sky provides an opportunity to make important and serious contributions to the science of astronomy. More importantly, since this is a hobby, it can be immensely enjoyable.

The observation and study of variable stars are as old as humankind and as fresh as the latest supernova.

Some reflection upon past variable star observations will serve as the beginning of our journey. For a few moments let us consider a nearby supernova that would have produced a brilliant star in ancient man's sky and how that event must have attracted his attention. Such an event could not have been ignored since nearby supernovae are very bright. We know that ancient humans observed these events because records of these bright stars have been found, for example, in Chinese and Native American records. In response to the sudden brightening of these cataclysmic events, history records the beginning of wars and the end of wars. Emperors of great kingdoms were crowned and some unlucky few were executed in response to the changes seen in stars. A complete understanding of how humankind's history has been shaped and altered as a result of the varying brightness of stars may never be known but it certainly can provide for some intellectual consideration. For example, a more recent stellar event occurred in 1572 when Tycho Brahe noticed a bright supernova in the constellation Cassiopeia. At such a late date in the history of humanity, this was the first time Western civilization became aware of variable stars. As a result, history followed a new path.

It took a few centuries longer, until the 1800s, before a German astronomer named Friedrich Wilhelm August Argelander (Figure 1.2) began the first, serious study of variable stars. Because of his efforts, F.W.A. Argelander is considered by some to be the father of variable-star observation.

In 1843 Argelander published a catalog of the stars as part of his study of variable stars that were visible to the naked eye and also created a unique method for estimating the brightness of the stars in relation to one another. His method for the estimation of the brightness of stars is called the *Argelander stepwise estimation method*.[1] Using his new catalog and his new method of comparison, within eleven years Argelander measured the position and brightness of 324,198 stars between +90° and −2° declination with his assistants Eduard Schönfeld and Aldalbert Krüger.

Subsequently, in 1863 Argelander published a catalog known as the *Bonner Durchmusterung*, abbreviated as "BD." In that same year Argelander became the founder of the Astronomical Society and together with

---

[1] This method of estimating the brightness of variable stars, among others, will be described later in this book.

Figure 1.2.
F.W.A. Argelander
(1799–1875).

Wilhelm Foerster, and others, began completing a survey of the celestial sky. In 1887 the society independently published a catalog of stars between 80° and –23° declination containing approximately 200,000 stars. This catalog is the *Astronomische Gesellschaft Katalog* (AGK). Argelander died on February 17, 1875, but his assistant Eduard Schönfeld extended the catalog by 133,659 stars within the zone ranging from –2° to –22° declination.

The southern sky was also mapped and, beginning in 1892, under the direction of J.M. Thome at the observatory of Cordoba in Argentina, 578,802 stars from declination –22° to –90° were collected as the *Cordoba Durchmusterung* (CD). This catalog was published in 1914. Together with the Bonner Durchmusterung, these catalogs built a compendium of more than one million stars down to 10th magnitude, a measure of star brightness explained later in the book.

Over the centuries, beginning with the effort of Argelander, the study of variable stars has secured the interest of many astronomers. Today, thousands of amateur astronomers from around the world observe and study variable stars. Some do it privately for personal satisfaction and intellectual curiosity while others belong to organized clubs or groups and conduct organized campaigns targeting hundreds of stars and collecting thousands of observations. As a result of the Internet, it is possible for amateur astronomers to share their observations with each other and in some cases with professional astronomers from around the world.

Most astronomers would agree that the study of variable stars is crucial to the overall effort of trying to understand the Universe. The assorted categories of variable stars represent stars in various stages of evolution. For example, the eruptive young T Tauri stars allow us to observe the birth of stars as they evolve from their protostar phase and enter adolescence. Supernova explosions bear witness to the violent death of gigantic stars as they produce as much energy in a few short seconds as our Sun will produce over its entire 10 billion year life. Binary stars, gravitationally bound in the orbits predicted by Johannes Kepler, provide a method of measuring a star's mass and allow us to judge the size of stars too distant to visit. Long-period pulsating stars named for the red giant that was first observed by the German astronomer David Fabricius in 1594, Mira – "the Wonderful," allow astrophysicists an opportunity to contemplate the interior workings of ancient stars. The energetic outbursts from dwarf novae furnish some of the best opportunities to study accretion disks and the underlying processes that may have contributed to the formation of our solar system and even the galaxies.

If you find the notion of personally participating in such a journey intriguing, you can hardly embark upon a more rewarding endeavor than the observation and study of variable stars. Come and join the company of astronomers such as Tycho Brahe, David Fabricius, F.W.A. Argelander and thousands of amateur astronomers from around the world. This is a journey of exploration in search for clues that help explain the workings of the cosmos. Come and explore the Universe with your own eyes. This is an invitation for you to become a participant in this great exploration and to play an important role in the history of variable-star observing.

Stellar Evolution

Since variable stars are stars, a basic understanding of how stars work will help you understand why variable stars are variable because not all variable stars are variable for the same reasons. By understanding why and how stars vary in brightness you may more easily select the type of variable star that you want to observe or study. The choice will be yours to make and an informed decision may allow you to avoid some serious frustration.

Consider the Mira-type variable star R Leonis that has a period of slightly more than 300 days. On the other hand, the delta Scuti-type variable star AZ Canis Majoris has a period of only 2 hours and 17 minutes. And the cataclysmic dwarf nova X Leonis goes into outburst about every 17 days. Interesting or confusing? Before you spend a year watching a star slowly vary in brightness or miss the opportunity to catch a much faster star in action, a little time spent understanding why variable stars act the way that they do may save you some valuable time later.

Using thermonuclear reactions, stars produce energy by converting light elements into heavy elements. Most stars convert hydrogen into helium to produce their energy. Hydrogen is the most abundant element in the Universe, and understandably, stars are composed mostly of hydrogen. It is believed by nearly all cosmologists that at the beginning of the Universe the only thing that existed besides energy was a lot of hydrogen, a little helium and a smidgen of deuterium and lithium. It's no wonder that stars are composed of hydrogen and helium if the early Universe was composed mostly of these two elements. Stars have been converting hydrogen into helium to produce energy since the beginning of time. But stars do not just convert hydrogen into helium. They convert helium into carbon and oxygen and nitrogen and eventually, even heavier elements. Except for the hydrogen and helium present at the beginning of the Universe, everything else is composed of the heavier elements that were formed over billions of years within the cores of countless stars. This process is called *nucleosynthesis*.

Over time, gravity brings hydrogen together into huge clouds that are light-years in diameter. You can go outside on any clear night, look into the sky and find these hydrogen clouds. The Orion nebula, the Trifid nebula and the region around Rho Ophiuchus, the Eagle nebula, the Tarantula nebula, the Cone nebula and the Lagoon nebula are a few of the places in the sky where you can easily observe these hydrogen clouds tonight with binoculars or a telescope. Our Solar System – the planets within, our Sun, you and me – is made from the stuff found within these huge hydrogen clouds. Carl Sagan called it "star stuff."

Eventually, still under the influence of gravity, the hydrogen cloud collapses even farther. After millions of years, the collapsing cloud crushes the hydrogen until its temperature and density are sufficiently high to cause the individual atoms to combine. Hydrogen atoms composed of one proton and one electron fuse to form helium atoms. This process is not as simple as it sounds. A wild dance that strips the electron from its partner and then forces the proton to combine with a free electron to become a neutron takes place. Eventually, proton–neutron pairs bond with a fast-moving free proton to form helium-3. In time, two helium-3 nuclei come together to form helium-4 and release two protons that return to the dance floor to look for new partners. This is a hot dance and it takes temperatures above a million degrees for all of this to happen. Relying upon this process, our Sun converts about 4 million tons of hydrogen to helium every second. It's been doing so for about 4.5 billion years and it will continue to do so for three or four billion years.

It was not until Albert Einstein explained that mass and energy are interchangeable[2] that astronomers understood this amazing process that powers the stars. If you come across a science book old enough, you will find that people once considered that coal may have been the source of the Sun's energy. We know better now, in part, thanks to Dr. Einstein.

After this wild dance that combines hydrogen atoms to form helium atoms, a little mass is lost from the original hydrogen atom. This lost mass is converted into energy and the force that initially brought the hydrogen atoms together, gravity, would collapse the hydrogen cloud into an incredibly tiny object except for this energy. Ultimately, the energy produced from the nuclear fusion of hydrogen exerts enough pressure to stop the complete gravitational collapse of the hydro-

---

[2] Einstein's famous equation, $E = mc^2$, that says energy is equivalent to mass.

gen cloud. When the inward collapsing hydrogen cloud is balanced by the outward energy pressure from nuclear reactions, a star is born. More importantly, if the outward pressure produced by the nuclear fusion exactly matches the inward gravitational collapse of the star, a stable star is born. However, not all stars are stable. Unstable stars pulsate, contracting and expanding in an attempt to find a balanced existence. During the convulsions that they experience in this attempt to find a harmonious balance within their lives, they vary in brightness; sometimes dramatically!

Most stars are using hydrogen as their source of energy and all of the stars that are using hydrogen for their energy source are related in a sense. It will be obvious once you start observing stars that not all stars are the same size, or the same temperature, or the same color. Some stars, like Betelgeuse in the constellation of Orion, and Mu Cephei, found in the constellation of Cepheus, are so big that they would fill a large portion of our solar system, gobbling up all of the inner planets and the asteroid belt and extending out even to the orbit of Saturn. Other stars like the white dwarf companion to Sirius are as small as the Earth, and neutron stars are only the size of a city. There are stars a hundred times hotter than our Sun and stars cool enough to contain water molecules. Some stars are deep blue, some are yellow and others are blood red. In contrast to this diversity, all stars that use hydrogen for their energy source are grouped together because of that one common characteristic. These are hydrogen-burning[3] stars and hydrogen-burners are called *main sequence stars*. Our Sun is a main sequence star because it is using hydrogen as its source of fuel.

Main sequence stars are usually stable stars although many vary in brightness. There is always an exception to everything but when talking about stellar evolution the main sequence is usually a good place to start since many of those stars are stable and most stars spend a large percentage of their lives as main sequence stars. Throughout this book the main sequence will be used as a point of departure when we investigate the different types of variable stars. It will be easier to understand the nature of the many different types of variable stars if we can start with a common reference point.

---

[3]"Burning" is a term used to mean thermonuclear reactions.

Now, back to the main sequence. As a star consumes its supply of hydrogen, a core composed of heavier elements forms. The production of these heavier elements starts with the fusion of hydrogen. Some of the older stars have converted a significant percentage of their original hydrogen into heavier elements and they are now using these heavier elements for energy. Eventually, if our understanding of the Universe is approximately correct, there will be no more hydrogen. It will all be converted into heavier elements and eventually those elements will be converted into even heavier elements.

As a star switches from using hydrogen as its primary source of energy to the heavier elements like helium, several changes occur within the star. First, it shrinks and gets hotter.

Remember, it takes temperatures above a million degrees to force hydrogen to fuse into helium but this is relatively cool for stars. At these high temperatures, with much of the hydrogen converted to helium, the star is beginning to grow hungry for a new energy source. The energy provided by hydrogen is dwindling but the star is not hot enough to cause helium to fuse. Helium fusion takes a temperature greater than 20 million degrees. The star is not hot enough so the helium is not providing any energy. It's just sitting there in the core of this hungry star. As the star begins its fast, it also begins to shrink; however, it doesn't lose weight. Gravity, the eternal force, has been patiently waiting, perhaps billions of years, for all of this to happen. Time is of no consequence to gravity and since the birth of this star, gravity has been waiting to compel the star's inward contraction. As the star cools, its thermonuclear pressure decreases and the star begins to collapse and shrink.

The star begins its collapse toward oblivion, but it also begins to get hotter, just as in the early stages of this star's birth when gravity first brought the hydrogen together and crushed it. Now, gravity is crushing the helium core, raising its temperature and density. Eventually, while the remaining hydrogen is still fusing within a thin shell surrounding the star's central region, the core reaches 20 million degrees. At just the right time, at just the right temperature and density, the helium dramatically flashes into a much hotter energy source for the star. The core of the star is now producing energy again but at much higher temperatures than when hydrogen was the source of fuel. This is the core of a helium-burning star and a helium-burning star is much hotter than a hydrogen-burning star.

The thermonuclear pressure from the hot helium core not only stops the inward contraction of the star, it actually pushes the outer atmosphere of the star outward. The star grows! It is larger in diameter now than when it was only fusing hydrogen. Its mass has not changed. Nothing new has been added. It's just larger because the much hotter core has pushed the outer edge of the star farther away, and as a result this outer edge is cooler than before. Because the outer atmosphere of the star is pushed farther away from the core, it has cooled somewhat. The core is hotter but the outer edge of the star is located farther away and as a result, it is cooler than before. Now the star has changed color. Because the outer atmosphere of the star is cooler it has become more red. Now, the star is evolving off the main sequence since it no longer uses hydrogen as its primary source of energy. As stars move off the main sequence, they are said to *evolve*. During this complex process, evolving stars move through evolutionary stages that have been given fascinating names such as the instability strip, the forbidden region, and the asymptotic giant branch. Perilous times await evolving stars.

The process described, to some approximation, continues as the star produces heavier elements and then in turn begins to use these heavier elements as an energy source. In many cases, evolving stars are unstable and a myriad of variable stars develop from these aging stars. The evolution of stars is an important part of the overall story of variable stars.

The Hertzsprung–Russell Diagram

Astronomy is interesting not only because it is a tool to explore the Universe but also because of the people involved with the exploration. The history of the world would not be nearly as exciting without Julius Caesar, Marco Polo, Leif Ericson, Christopher Columbus, Jules Verne or Neil Armstrong. Much of the history of the world is the story of exploration and the history of astronomy is also a story of exploration. In essence, you

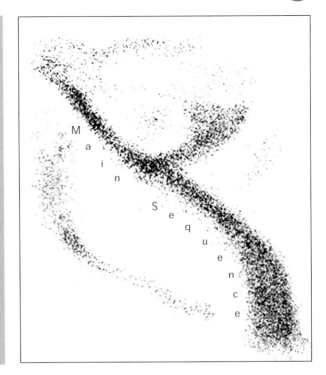

Figure 1.3. The Hertzsprung–Russell diagram showing the main sequence as a dense line of stars running from the upper left to the lower right.

are interested in astronomy because you are also an explorer.

Two early explorers, in the astronomical sense, developed what is considered the cornerstone for the understanding of stellar evolution, known as the *Hertzsprung-Russell diagram* (Figure 1.3). Stellar evolution is of vital importance when considering variable stars and so we will introduce the Hertzsprung–Russell diagram here. It is sometimes called the HR diagram for short.

In the early years of the twentieth century, Ejnar Hertzsprung was a Danish astronomer but interestingly had no formal education in astronomy. He was in fact an amateur astronomer. Professionally, he was a chemical engineer. Because of his formal education he was keenly interested in the chemistry of photography and turned to astronomy in 1902 when he began working in small Danish observatories. Using his understanding of photography to measure starlight he was able to show a relationship between the color of stars and their brightness.

During the time that he spent in various observatories in Denmark, Hertzsprung's work eventually came to the attention of Karl Schwartzschild[4] who was the director of the Potsdam Observatory at the time. As a result, in 1909 Schwartzschild soon found a place for Hertzsprung on the staff of the Göttingen Observatory where he eventually became a senior astronomer.

Meanwhile, at about the same time that Ejnar Hertzsprung was working in Europe on classifying stars based on their spectra, Henry Norris Russell was doing the same thing in America. It is said that he was shown the transit of Venus across the Sun's disk when he was five years old and that this may have sparked his interest in astronomy. In any case, many years later, after receiving his Ph.D. in astronomy at Princeton, he left to work at the university observatory at Cambridge. Along with possessing a great interest in binary stars, Russell was interested in stellar spectra and eventually concluded that there were two main classes of stars, one much brighter than the other. In due time, he was able to show a relationship between the true brightness of a star and its spectrum.

This is all very interesting but the fascinating thing is that Hertzsprung and Russell, working independently, were able to show a color and temperature relationship between stars. In other words, the color of a star will indicate its temperature to a close approximation. It is on the Hertzsprung–Russell diagram, named in their honor, that the main sequence stars are positioned as a roughly diagonal line extending from the top left to the lower right. The Hertzsprung–Russell diagram allows astronomers to plot stars according to their temperature, color, luminosity and age.

If you are hunting for variable stars, there are well-known locations within the HR diagram where variable stars can be easily found. In other locations on the HR diagram, variable stars hide within a multitude of stable stars and are difficult to locate. Understanding the evolution of stars and how the HR diagram can be used to track their evolution will aid you in a deeper understanding of the characteristics of stars. The HR diagram used as a common reference point, along with a deeper discussion of stellar evolution, will be used in later chapters to assist you in preparing to understand and observe variable stars.

---

[4]Karl Schwartzschild is known for laying the foundation of the theory of black holes, demonstrating that bodies of sufficient mass would have an escape velocity exceeding the speed of light. In one of the many tragedies of World War I, Schwartzschild died of illness in the trenches while fighting as an infantry soldier.

A Profile of Variable Stars

So what are variable stars?

Very simply stated, variable stars are stars that vary in brightness. This variability can occur over the span of seconds as in the case of ZZ Ceti stars, it can take years as with Mira-type variable stars or it can happen within several hours and just once as in the case of a *supernova*.

Each of the variable stars just mentioned vary in brightness for different reasons. The ZZ Ceti stars are pulsating white dwarfs that change their brightness with periods that range from approximately 100 to 1000 seconds. Occasionally, flares occur and will result in the star doubling its brightness.

The Mira-type variable stars are supergiant or giant stars that also pulsate but instead of within seconds, the measured period of their pulsation lasts hundreds or even thousands of days. These stars are of interest to astrophysicists because they are representatives of a very short-lived phase of stellar evolution. Mira stars are ending their lives and some will soon become planetary nebulae. Because they are so large and their atmospheres so far removed from the core, their extended atmosphere is extremely rarefied and would be considered an excellent vacuum here on Earth.

The rare supernova dramatically alters the structure of a star in such a way that the star is irrevocably changed. The shell of the star is violently expelled and eventually interacts with the interstellar medium, forming a supernova remnant. It is possible to observe many supernova remnants in the night sky and they serve as a reminder that the Universe is constantly changing. Supernovae could be thought of as the recycling facilities of the cosmos. Almost everything comes from supernovae and almost everything will go back into producing new supernovae.

These examples are just a few types of variable stars and many other types of variable stars exist. Currently, astronomers recognize over 80 types of variable stars, an additional five types for eclipsing binary variables according to the companion's physical characteristics, nine types based on the degree of filling of inner Roche lobes[5], and ten types of optically variable close binary sources of strong, variable X-ray radiation. The number of possible combinations is staggering and as a result, variable-star observers usually specialize in just a handful of types. Some observers spend a few years observing a half-dozen classes of variable stars, then move on to observe different types. Other variable-star observers observe only a few classes of stars, never moving to other types. As a result, they are very knowledgeable after having spent years or even decades specializing on their selected stars.

Finding a variable star is not too difficult to do. Because all stars oscillate to some degree, they are all to a greater or lesser degree variable, and these oscillations give rise to variations in luminosity. However, the amplitude of this oscillation for any given star is usually very small, so that the associated variations in luminosity are tiny. Not very exciting since these micro-variations are invisible to the naked eye.

For example, solar-type stars vary in luminosity by an order of micro-magnitudes, far too faint to be observed with the naked eye. The good news is that you can study these micro-oscillations with instruments such as stellar photometers and charge-coupled devices that are available to amateur astronomers. The study of small-amplitude stellar oscillations has become an important branch of astronomy because the oscillations involve the entire star, so from them one can glean information about the deep stellar interior. An excellent example of this type of effort is the study of Delta (δ) Scuti stars.

Some stars vary in brightness because of intrinsic properties but others vary in brightness as a result of external characteristics that have nothing to do with the actual make-up of the star. For example, two stars that would not be considered variable individually may be classed as variable stars if they orbit each other in such a way that one moves in front of another and causes an eclipse. In some cases, this eclipse will result in a reduction of luminosity when viewed from Earth. In effect, the two stars vary in brightness. This variability is not a result of the inherent properties of either individual star but rather is a result of their interaction with each other. Now, imagine two different classes of variable stars that are each variable for intrinsic reasons, perhaps one is pulsating while the other

---

[5]Roche lobes will be explained in Chapter 5, Cataclysmic Variables.

experiences flare activity, orbiting each other in such a way as to cause an eclipse when observed from Earth. Understanding this complex type of configuration can be difficult at best but the labor required to understand this type of situation is just part of the challenge of variable-star observing. Six major classes of variable stars will be described to show the diverse mechanisms for producing variability.

Eruptive variable stars are defined as stars that show a sudden, large outburst of energy causing their visual brightness to increase by 200 times or more in a few days. These outbursts of energy are caused by violent processes, such as flares, that occur within the visible edge of the star. In some cases, stellar material is being blown away from the star and interacts with the surrounding interstellar medium, causing changes in visual brightness.

Pulsating variable stars show a periodic expansion and contraction of their surface layers, as if breathing. In some cases, the expansion occurs uniformly throughout the star. In other cases, the star quivers during unequal expansions that occur within the various layers of the star.

Cataclysmic variables show outbursts caused by thermonuclear processes on or within their surface layers or deep within their interiors. Novae and dwarf novae are members of this much-observed class of variable star and an important property that is shared by these stars is that they are all extremely close binary systems, in most cases possessing orbital periods of less than half a day. Supernovae are recognized as cataclysmic variable stars too. The first supernova to be observed using modern techniques was the variable star S Andromedae in the Great Andromeda galaxy, also known as M31. When it was first observed in 1885, it was believed to be a normal nova, with a relatively modest brightness. This led to estimates of the distance to M31 that misled astronomers about the size of galaxies. In 1924, Edwin Hubble discovered a type of variable star known as a Cepheid in M31 and was able to make an independent estimate of the distance using the *period–luminosity relationship*. It then became clear that S Andromedae was more than ten thousand times brighter than an ordinary nova. Supernovae are rare events, with a typical galaxy producing no more than about two or three per century.

Rotating variable stars possess irregular surface brightness and/or elliptical shapes. Their variability is caused by axial rotation with respect to the observer. The irregular surface brightness may be caused by the presence of spots or by some thermal or chemical variations of the atmosphere caused by magnetic fields. Our Sun displays sunspots that are the size of the Earth. Imagine a star with sunspots that are the size of our Sun!

Eclipsing binary stars are multiple star systems that have the orbital plane of the orbiting stars oriented approximately along the line of sight of the observer so that one star may periodically pass in front of the other thereby blocking the light of the eclipsed star. The study of these light curves not only reveal the presence of two stars but can also provide information about relative temperatures and radii of each component from the amount of light decrease and the length of the eclipse. Recently, amateur astronomers have detected the possible transit of extra-solar planets across distant stars.

Optically variable X-ray sources are a somewhat ambiguous class of variable stars. Some astronomers consider X-ray binaries to be any kind of interacting close binary with a compact degenerate object, such as a white dwarf, a neutron star, or a black hole. The definition that we will use is that X-ray binaries are only those interacting close binary systems that contain a neutron star or black hole. The main difference between the cataclysmic variables and the X-ray binaries is the X-ray luminosity. Many X-ray binaries produce optically variable light phenomena that can be observed by amateur astronomers. Another type of high-energy variable star is the *gamma ray burst*, also known as a GRB. These enigmatic objects have just recently come to the attention of amateur astronomers. Presumably stars, these objects are so far distant that we have found no evidence of their existence on any deep image taken before their outburst. They eventually fade away and again nothing is seen where the GRB appeared. Today, amateur astronomers can be notified through a network of interested astronomers when satellites have detected a GRB. If quick enough, amateur astronomers have been able to catch a glimpse of these interesting objects.

These six classes of variable stars; eruptive, pulsating, cataclysmic, rotating, eclipsing, and variable X-ray sources will be characterized in the subsequent chapters of this book. Suggestions of how best to observe each class of star and the basic methods for recording and analyzing your observations will be given so that you may preserve and later examine the fruits of your labor. A description of the basic telescope and binocular types that variable star observers use, as well as the various eyepieces, mounts and other accessories that you will routinely use to observe variable stars will also be explained. Along with visual observation techniques, the use of semiconductor light detectors (known as CCDs), photoelectric (PEP) methods and the use of science filters will be briefly investigated.

Chapter 2

The Variable Stars

Perplexity is the beginning of knowledge.

Kahlil Gibran

Recalling that variable stars are a particular type of star, we'll return to the discussion of stars in general and eventually how variable stars receive their names. Before you begin to observe variable stars you really should have a proper introduction.

It will be a little easier to get to know variable stars if we can categorize them somehow. Presently, there are over 36,000 variable stars contained within the *Combined General Catalog of Variable Stars* (*GCVS*) and unless we figure out a way to lump most of these stars together into small groups, you're never going to be able to study them in a systematic manner. And the *GCVS* is not the only inventory of variable stars! When you begin to consider the HIPPARCOS variable star annex, the All-Sky Survey, the ROTSE variable star survey, the MACHO project, the MISAO project and the STARE project, to name just a few, the number of variable stars can far exceed the number reported within the *GCVS*. A systematic method of nomenclature is required.

Let us gently advance our understanding of variable-star classification. We'll begin with stars in general. They certainly are puzzling and perplexing but there are important characteristics of these interesting objects that will quickly become apparent to you as an astronomer. One of these characteristics is brightness.

When you begin to observe the stars, you'll soon notice that not all stars are the same brightness. Some are brighter than others and some are so faint that you will not be able to see them without a telescope. Most stars are so faint that you will never see them. The definition of a star's brightness is called its *luminosity* and this is a measurement of the total energy emitted by the star each second and much of this energy is invisible to your eye. The expression of brightness, *the light that we can see*, is made using the term *magnitude*. When you observe variable stars and wish to specify stars that have different brightnesses than others, you will relate their respective brightness by comparing their magnitude. Comparing the brightness of stars is fundamental to variable star observing. In every instance you will be comparing the brightness of one star, the variable, to another, the comparison star, so as to make an estimate of the variable's magnitude.

The brightest stars are described as having a magnitude of one. This is a measure of their brightness, and magnitude one stars are some of the brightest stars in the sky. A few of the very brightest stars, such as Sirius and Rigel, have a magnitude closer to zero, while the fainter stars have magnitudes that are indicated by larger numbers. Generally, magnitude six stars are the faintest stars visible to the unaided eye when viewed under exceptionally dark skies. You will need binoculars or a telescope to observe stars fainter than magnitude six, for example magnitude seven or magnitude eight stars.

The ancient astronomers who worked without the aid of telescopes named the brighter stars. These ancient names are still in use today. If you grab a sky atlas and examine the list of star names you will find bright stars with unusual names such as Zubenelgenubi, Betelgeuse, Denebola, Vindemiatrix, Kornephoros, Menkalinan and Pherkab. Sadly, not all stars have a formal name such as these enchanting appellations. Instead, the Greek alphabet was used to designate many stars and many are still designated using the Greek alphabet from brightest to faintest. The brightest star in a constellation is usually designated *alpha*, followed by *beta*, *gamma*, *delta*, and so forth through the Greek alphabet. Betelgeuse is known as alpha (α) Orionis since it was believed to be the brightest star in Orion when it was named. It has since been found to be a variable star and a little fainter than Rigel, beta (β) Orionis (also a variable star). Such mistakes are common and in general, not remarkable.

Another characteristic of the stars that you will notice is color. Stars have color and when observed through binoculars or a telescope the colors can be spectacular. The very hot stars are blue and the cool stars are red. Between these two extremes in color you will find blue-white, white, yellow, and orange stars. Remember how the HR diagram shows a color-temperature relationship? Now we can look at that relationship a little closer because temperature and color are important to a variable-star observer.

Temperature is an important characteristic used to group stars and today stars are categorized by temperature as well as color. However, neither temperature nor color were the first characteristics used to scientifically classify stars. When stars were first being arranged so that common physical characteristics could be identified and used to group similar stars together, their spectra were selected to be the primary common characteristic.

A star's spectrum shows its chemical composition. Our star, the Sun, shows a rich spectrum that serves as a record of the various elements found within it. Helium, the second element on the periodic table of elements and one of the primordial elements present at the birth of the Universe, was first discovered in our Sun when its spectrum was examined closely. Later, helium's existence was confirmed when it was found on Earth. As we'll see, the spectrum of a star depends upon its temperature.

This early nomenclature, based upon stellar spectra, began with stars classified as A-type stars and progressed through the alphabet. The beginning of the alphabet was a logical place to start. All A-type stars had a similar spectrum as did the B-type and C-type stars, advancing through each of the letters. However, it didn't take long before astronomers noticed that the stellar classification scheme beginning with the stars classified as A-type and running through the alphabet, using each letter, was flawed. This progression did not proceed as hoped. A-type stars didn't blend into B-type stars and they, in turn, didn't blend into C-type stars. Something was wrong (Figure 2.1).

Astronomers began to understand that if the classification was rearranged by moving some of the stars to different positions within the sequence then one group of stars neatly blended into the next group. The O-type stars blended into the B-types stars which then blended into the A-type stars and so forth. Instead of beginning the classification with the letter A and then progressing through the alphabet, this reorganization

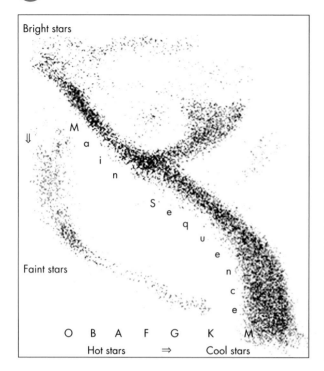

Figure 2.1. The Hertzsprung–Russell diagram showing the main sequence with the change in luminosity and temperature indicated.

resulted in stars already classified as O-type stars being placed first in order, followed by stars previously classified as B-, A-, F-, G-, K-, and M-type stars.

It was obvious that this arrangement allows a temperature sequence to be represented with the hottest stars listed first with cooler stars following. Within the HR diagram, the hot stars are positioned to the left and the cooler stars are positioned to the right. This whole system is known as the *spectral classification* system and it can be remembered by using this mnemonic: *Oh, Be A Fine Girl (Guy) Kiss Me*. Several other classes of stars have been added since this classification method was developed and we'll discuss those classes of stars later in the book. For right now, O- through M-type stars will suffice to adequately explain the HR diagram.

After adopting this system, another problem soon came to the attention of astronomers. Simply using one letter to classify all stars of a similar spectral type proved to be restrictive, and so numbers within each letter class were added. Now the sequence not only runs from O-type stars through M-type stars but within each letter classification ten numbers beginning with 0 and progressing through 9, with a few exceptions, are used to further define each classification. This method of stellar classification begins at about O3 and advances to O9[1] and then changes to B0 and continues to B9. After B9 comes A0 and so forth through F, G, K and M stars. These first few classes of stars are very hot stars and are referred to as *early* stars. As this classification scheme moves through the F-type and on to the M-type, the stars are referred to, relative to earlier stars, as *late* stars. Also, within a single letter classification, a zero star is considered earlier than a nine star. For example, an A0-type star is earlier than an A7-type star. You may also say that an A7-type star is later than an A0-star. It all depends upon your point of reference.

Used in conjunction with the spectral classification just described, there is also a luminosity classification. Hertzsprung's and Russell's greatest discovery, made in the early twentieth century, was that many stars do not lie on the main sequence. The principal additional feature of the HR diagram is a central band that goes up and to the right, away from the main sequence, in which luminosity increases as temperature decreases. To be both bright and cool requires great size. These stars are therefore naturally called *giants*, discriminated by calling those stars of the main sequence *dwarfs*. In spite of the size of bright main sequence stars, the terms "main sequence" and "dwarf" are synonymous. In comparison to our star, giant stars can easily encompass the inner Solar System. We also find stars below the main sequence – stars that are both hot and quite faint. These must be terribly small, even smaller than the Earth itself. The first stars of this type found were white, so the name white dwarf was applied, a term still used even though some of these stars are red and others blue.

These various stellar zones on the HR diagram were formalized in the 1940s by astronomers W.W. Morgan, P.C. Keenan, and E. Kellman, who placed them into luminosity classes distinguished by Roman numerals I through V, representing supergiants, bright giants, giants, subgiants, and main sequence dwarfs respectively. Keenan also suggested a class for stars larger and brighter than the supergiants – the "hypergiants." These grand stars are designated luminosity class "0" which can lead to some confusion with the decimal spectral classes and the letter "O." The zero class of star

---

[1] The spectral class of O-type stars does not use ten numbers. This class begins with 3 and progresses through 9.

was first applied to the S Doradus stars. For example, the chaotic star η (Eta) Carinae is a hypergiant, spectral class B0 0, and possesses a mass equal to approximately one hundred suns. This luminosity classification is called the MKK or the MK system. All of this is important when you begin to study variable stars because, in essence, the luminosity classification system describes the size of stars. References to the luminosity classes of stars will be made throughout the book. As you begin to use this terminology, you will better understand it.

Now, back to the color of stars. Some names for stars are directly related to their color. Antares, the bright red supergiant that is sometimes called the "Heart of the Scorpion" because of its position within the constellation of Scorpius, received its name because it was considered to have a color similar to Mars. In Greek mythology Mars, the Roman god of war, is known as Ares. The name Antares is derived from *ant-Ares*, the rival of Aries. Aldebaran, the orange K-type star found within the constellation of Taurus the bull, is known as the "Eye of the Bull," probably as a result of its color as much as its position. Other examples also exist.

Of all the stellar characteristics that are observable, the one that will become most important to you as a variable-star observer will, of course, be that some stars vary in brightness sufficiently for you to measure. As we discussed earlier, variability can be a result of the physical structure of the star or as a result of how two or more stars block the light that we receive from them. It would seem, on first examination, that simply classifying variable stars by the mechanism that is responsible for their variability would be the best and easiest method. It certainly seems sensible that there can only be a few reasons that stars vary, so we will explore this idea later in this chapter.

Interestingly, some stars may even have been named as a result of their variability. Algol, the bright blue-white star found within Perseus, derives its name from the Arabic "Head of the Demon." Other descriptions for Algol include Head of Medusa, Eye of Medusa and the Ghoul Star. Perhaps ancient people so named this star because its light variations suggested something monstrous or evil. Whatever the reason, the names of stars provide interesting and descriptive insights into humanity's past. On those nights when observing is impossible because of clouds or wind, you can still explore the Universe through the observations and interpretation of ancient astronomers.

All stars, even those with formal names or those designated using Greek letters, also have modern catalog numbers. It seems that there are as many catalogs as there are stars to place into those catalogs. While examining a star atlas or catalog, you will find designations for stars such as SAO, HD, HR, HIP and *GCVS*, to name a very few. These catalog designations allow astronomers to indicate a star using a method that uniquely identifies that one star, even though one star may be found in many catalogs. As you examine your star atlas, a star catalog or use a computer program that displays the sky with information regarding each star, you will see these various catalog names. For example, the bright $2^{m}.8$ star[2] Zubenelgenubi, found within the constellation of Libra, is also known as SAO 158840 within the *Smithsonian Astrophysical Observatory* catalog. Within the *Henry Draper* catalog it is known as HD 130841 and when found within the *Hipparcos* catalog it is designated as HIP 72622. Regardless of whether you call it SAO 158840, HD 130841, HIP 72622 or Zubenelgenubi, you are in all cases referring to one, unique star.

Variable stars have their own catalog too. It is called the *Combined General Catalog of Variable Stars*, designated as the *GCVS*. For those stars that seem to be variable but have not been studied well, there is the *New Catalog of Suspected Variable Stars,* designated the *NSV*. The star Zubenelgenubi is found within the *NSV* and is designated as NSV 06827.

While traveling the road taken by those tenacious souls known collectively as variable-star observers, you will become familiar with various star catalogs, especially the *GCVS* and the *NSV*. The names of variable stars, their locations and their fascinating characteristics will be of paramount interest to you.

Those considered experts when it comes to the classification of variable stars are well aware of the difficulty encountered when constructing an accurate and consistent catalog of the various classes and types of variable stars. As early as 1880, the astronomer Edward Pickering made one of the first attempts at classifying variable stars. In his original classification scheme, he recognized five classes of variable stars: 1) new stars (*novae*), 2) long-period variables,

---

[2]Throughout the rest of the book, we will display the brightness of a star by its magnitude, in this way: $2^{m}.8$, meaning that the star has a magnitude of 2.8.

3) irregular variables, 4) short-period variables and 5) eclipsing variables. As time eventually revealed, this system had flaws and as the number of variable stars increased, its failings became more apparent. Later, in the early years of the twentieth century, Henry Norris Russell made another attempt at classification of variable stars. In his scheme, he also described five classes of variable stars: 1) novae, 2) long-period variables, 3) irregular variables, 4) Cepheids, and 5) eclipsing stars. Again, the classification system for variable stars was found to be less than perfect. In part because stellar pulsation was not known to be a mechanism for stellar variability, these early classification schemes were doomed to fail. Astronomers needed a good theory that would describe the energy source that powered the stars before a comprehensive understanding of variability could even begin.

As 1925 approached, aided by the great breakthroughs in nuclear physics taking place in the early years of the new century, astronomers began to understand the internal workings of stars. In particular, two more recent advances helped astronomers visualize variable stars within their new understanding of stellar development: the application of nuclear physics regarding the source of stellar energy and the development of sophisticated computer techniques. Computer techniques here mean advanced mathematical methodology used by assistants who where usually required to perform the complicated and time-consuming mathematical analysis. These assistants, usually women with advanced degrees in mathematics, where known as "computers."

With the acceptance of atomic theory, astronomers had their energy source for the stars, and variable-star classification blossomed. Still, there were problems with the classification of variable stars as the number and variety of these stars continued to increase. Astronomers struggled for years, attempting to place variable stars within a few well-defined and unique classes that adequately described the myriad characteristics discovered as their numbers increased.

Finally, in 1960, firmly established by the recommendations of the International Astronomical Union made in Moscow two years prior, the first classification outline was published. This classification outline has frequently been reworked and extended during the past 40 years. At its heart, and a principal source of information on the latest classification of variable stars, is the *Combined General Catalog of Variable Stars.* The *GCVS* contains classification and miscellaneous data on over 36,000 variable stars that have been assigned variable-star names. This catalog is considered the authority for the systematic placement for variable-star knowledge. Supplements to the *GCVS*, the well-known *Name List of Variable Stars* (*NL*), appear regularly in the *Information Bulletin on Variable Stars* (*IBVS*). In addition, *The New Catalog of Suspected Variables* (*NSV*) contains data on almost 15,000 new unnamed variable objects. Amateur astronomers who use the *GCVS* and *NSV* must understand that these catalogs are not just a compilation but that they are based upon a critical evaluation of the underlying data. However, errors exist.

If all of this sounds a bit confusing, consider the comments found in a paper written in 1978 by Cecilia Payne-Gaposchkin, Center for Astrophysics, entitled *The Development of our Knowledge of Variable Stars.* She says that "the classification of variable stars has undergone a change that recalls the supersession of the Linnaean system by the modern system of botanical classification." Even professional astronomers can find aspects of variable-star classification a bit confounding at times.

Along your journey you will, without a doubt, familiarize yourself with these catalogs and lists. Soon you will notice that almost every newly published catalog or review paper introduces new classes or subclasses of variable stars. Observers are well aware of the continual changes regarding variable stars. They understand that it is a necessary process that not only clarifies the many classes of variable stars but can also be confusing because, occasionally, it also becomes less consistent. It is not unusual to find a star listed within different classes of variable stars, or to see the transfer of a star from one class to another, and then back as time passes. These changes reflect the dynamics of the field. They are not only due to more and more sophisticated detection techniques, but also to the increasing time baseline along which data are being collected.

Another aspect of published classification schemes is the pronounced vagueness with which some classes are being defined. The *GCVS* lists several subclasses in which variable stars are characterized by the label *poorly studied.* Eventually this will result in better refinement and to the definition of new classes and subclasses of variable stars but for now it can be confusing. For example, within just the last few years we have seen several new classes of variable stars evolve like the rapidly oscillating Ap stars, the slowly pulsating B stars, γ Doradus stars, RPHS stars and EP stars. These types of stars have always been variable but only recently have they been officially recognized and new classes of variable stars are in the process of being compiled such as the λ Boötes stars, TOADs, and Maia stars. Of course, some of these potentially new classes of variable stars will probably be still-born, but the process continues.

In many cases, the actual classes of variable stars do not have clear-cut boundaries. The distinction between the various classes of variable stars is found within the characteristic that is being used to make the classification. If that characteristic is something that can be empirically measured, such as spectral type, then clear borders can be assigned. However, if the parameter is a physical quantity, for example mass, then a certain amount of doubt results. This happens not only because mass is derived from other observations with their own uncertainties that lead to errors, but also because the mass range forms an uninterrupted scale that is not neatly segregated or sharply defined.

By tradition, variable stars are classified into two main families: intrinsic variable stars and extrinsic variable stars. Intrinsic variable stars vary due to physical processes within the star itself, and extrinsic variable stars vary due to processes external to the star, such as rotation. Extrinsic variables are eclipsing binaries and rotational variables. Intrinsic variables are pulsating variables, eruptive and explosive variables. Among the intrinsic variables, the most difficult to classify are the explosive and symbiotic variables.

In the subsequent chapters of this book I will consistently indicate a class of variable star either by the abbreviated name or by the full name, for example, either W UMa type variables, or W Ursae Majoris stars. For the variables themselves, I will use the abbreviated *GCVS* notation, for example R And (a long-period Mira-type variable) when available. When a *GCVS* name is not available, a well-recognized catalog name will be used such as SAO or HD.

The remainder of this chapter explains the classification schemes of the *GCVS* along with an explanation of how this book will either follow the *GCVS* scheme, or deviate from it. When we deviate, I am not attempting to suggest changes to the classification scheme. Deviations will be made only to facilitate the understanding of the class as a whole, or to include within a class, variable stars that are not officially recognized by the *GCVS*.

Presently, the International Astronomical Union (IAU) is responsible for naming variable stars. The names are given in the order in which the variable stars are discovered within a constellation. If a star already possessing a Greek letter name is discovered to be variable, then the star will be referred to by the Greek name. For example, alpha Scorpii (α Sco), also known as Antares, is allowed to keep its Greek letter designation that indicates that it is the brightest star within the constellation of Scorpius. Otherwise, the first variable star discovered within a constellation is given the designation using the letter R, the next S, and so on to the letter Z. The reason for this unusual sequence will soon be explained.

Continuing after Z we return to R and begin double letter designations so that the next star is named RR, then RS, and so on to RZ. Then we begin again with SS to SZ, and so on to ZZ. Then the naming starts over at the beginning of the alphabet: AA, AB, and continuing on to QZ. This system, where the letter J is always omitted, can accommodate 334 names. However, there are so many variables in some constellations that an additional nomenclature is necessary. After QZ, a variable star is named V335 since 334 variables have already been named. The names continue as V336, V337, V338 and so forth. The letters representing stars are then combined with the possessive Latin form of the constellation name in the same way that the Greek alphabet is used for complete identification of the variable star. Examples are SS Cygni (SS Cyg), AZ Ursae Majoris (AZ UMa), and V338 Cephei (V338 Cep).

Friedrich Argelander established this system of nomenclature. He started with a capitalized R for two reasons: the lowercase letters and the first part of the alphabet in capital letters had already been allocated for other designations, leaving capitals towards the end of the alphabet mostly unused. Argelander also believed that stellar variability was a rare phenomenon and that no more than nine variables would be discovered in any constellation. The letter J is always omitted. The reason for this is sometimes regarded as a mystery, lost in the dusty annals of astronomical history. The reason is simply so as not to confuse the letters I and J.

The *American Association of Variable Star Observers* (AAVSO) uses a second system of names – a numerical designation. This numerical designation, named after the Harvard College Observatory, where the system was first used, is a group of six numbers and a sign that gives the variable's approximate coordinates for the year 1900. The first four digits give the approximate hour and minutes of right ascension, the last two prefixed with a plus or minus sign, the degrees of declination. For example, the designation 0942+11 for R Leonis indicates an approximate position of right ascension of 09 hours 42 minutes and a declination of +11 degrees for the year 1900.

The Classification of Variable Stars

As you are beginning to understand, there are myriad types of variable stars but they can generally all be arranged into a handful of reasonably well-defined classes: the six classes described in Chapter 1. Of course, new classes of variable stars will inevitably be suggested and some stars will change designations in the years to come. Perhaps your observations will be responsible for some of these changes. Certainly, changes will have occurred in the short time since the publication of this book.

In this light, I am aware that a few astronomers will argue that a star or two may be misplaced within the following arrangement, especially those stars not officially recognized within the *GCVS*. My humble intention is to keep our examination of the major classifications as simple as possible without violating the currently accepted nomenclature and with due consideration for the dynamic criteria with which variable stars are judged, labeled and classified. With that said, I fully understanding that a zestful debate, conducted elsewhere, regarding the proper classification of some of these stars could be considered enjoyable, perhaps even obligatory.

By keeping the classification scheme as simple as possible, it should be a little easier to learn the numerous types of variable stars; however, as a precaution against unrealistic expectations, I refer you back to Kahlil Gibran's quotation at the beginning of

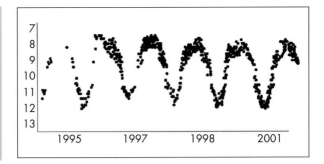

Figure 2.2. Light curve of the Mira-type variable star, T Cas. This type of star shows a periodic variation in brightness. Data provided by VSNET. Used with permission.

this chapter. Fortunately the six major classifications of variable stars are based upon some obvious characteristic of the star or upon some characteristic (shape or *morphology*) of the light curve produced by plotting the star's variation in luminosity. We'll be closely examining these characteristics within the book (Figures 2.2 and 2.3).

I want to mention *light curves* early and explain them later[3] because it will be a descriptive term that we use throughout the book. A light curve is a two-dimensional plot[4] of a variable star's changing brightness analogous to plotting the change in temperature from season to season or the rise and fall of stock prices. The amazing thing about light curves is that they can be diagnostic. In other words, in some cases the morphology (shape) of a variable star's light curve is sufficient to determine the type of variable star that produced it. Be aware that this is *not* the case with every variable star

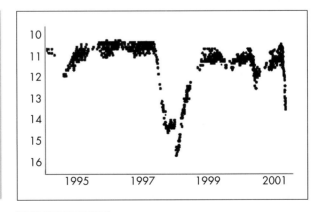

Figure 2.3. Light curve of the RCB-type variable star, SV Sge. This type of star does not show a strictly periodic variation in brightness. Data provided by VSNET. Used with permission.

[3] Light curves will be explained at length in Chapter 13.
[4] A two-dimensional plot is also known as a graph. Light curves are graphed using scatter or line graphs. We will construct light curves later in the book.

and every light curve. Most light curves are ambiguous and you must be very clever to figure out the type of star or stars that may be indicated. We will examine light curves, their construction and their analysis, later in the book. Now, it's time to examine the six major classes of variable stars.

In the next six chapters we'll examine eruptive variables, pulsating variables, cataclysmic variables, rotating variables, eclipsing binary systems and optically variable X-ray sources using the *Combined General Catalog of Variable Stars* as a guide. Beginning each chapter you will notice the description of each class of variable star taken directly from the *General Catalog of Variable Stars*. The description is verbatim. A short introduction to each group followed by a table listing their official or unofficial names is then presented. A brief description of each group is then provided. Old or seldom used descriptions and names of variable stars are also explained. When the *GCVS* is referenced, it is assumed to mean the *Combined General Catalog of Variable Stars* that includes the Name-lists 63 through 75, unless otherwise indicated.

An observation key can be found next to each class of variable star, within the margin. This key will indicate the general brightness, amplitude limits, period range and best observing technique for each class of star. When a large number of stars within a class can be found brighter than $10^m_.0$, the key will indicate "Bright stars." When the amplitude limits within a class of variable star are generally less than $1^m_.0$, the key will indicate "Small amplitudes." When the periods displayed within a class of variable star are generally longer than a day (24 hours), the key will indicate "Long periods." You can certainly deduce the converse. And finally, the best method for observing each class of star will be indicated as visual, CCD or PEP (CCD and PEP methods will be explained within Chapter 12). In many cases, any observation method will work. In other cases, one method may work for one particular reason, such as detecting a dwarf nova outburst or a supernova, and another method will work for a different reason, such as measuring small changes in brightness. Many variable stars can be observed using all three methods. In any case, the observation keys are meant to be used as quick guides. As with most of the subject matter within this book, there will be exceptions.

When the brightness of a star is described, it will usually be a measurement of the visible light and may be indicated as "V." For example, the variable star KU And has an amplitude ranging between $6^{m}\!.5$ and $10^{m}\!.5$ in V. Several color bands within the spectrum are used by astronomers, including ultraviolet (U), blue (B), visual (V), red (R) and infrared (I). If the color band is not defined, assume that it is in V.

Chapter 3

Eruptive Variable Stars

Eruptive variables are stars varying in brightness because of violent processes and flares occurring in their chromosphere[1] and coronae.[2] The light changes are usually accompanied by shell events or mass outflow in the form of stellar winds or variable intensity and/or by interactions with the surrounding interstellar medium.

GCVS

There are over 3900 eruptive variable stars within the *General Catalog of Variable Stars*. Of this number, over 550 are classified as *uncertain* within the various groups. Variable stars are identified as uncertain when a colon (:) follows the variable star type (e.g. FU: or RCB:) and indicates that insufficient information exists to definitively identify the variability type. It means that additional data must be collected and analyzed so that the star can be properly categorized.

After inspecting the individual characteristics of each type of star that comprise this class, the singular classification of eruptive stars is arguably the most tangled. There is no single mechanism responsible for causing the characteristic eruptions of the class as a whole. Unlike pulsating stars, rotating stars or eclipsing binary stars, in which each class exhibits variability attributed to a single, broadly defined and certainly complex mechanism, eruptive variable stars demonstrate light variability because of ill-defined, unique, or unrelated reasons. Even cataclysmic variable stars, with

---

[1] The layer of gases above the visible edge (photosphere) of a star in which sunspots, flares and prominences occur.
[2] The thin outer atmosphere of a star.

many different mechanisms responsible for their behavior, are at least all … cataclysmic! For example, within this group of variable stars FU Orionis stars are variable because they are releasing gravitational energy, γ Cassiopeia stars display variability due to a hot shell or cloud surrounding the star, R Coronae Borealis stars are both eruptive and pulsating stars, RS Canum Venaticorum stars are chromospherically active, UV Ceti stars display flare activity and the Wolf–Rayet (pronounced rye-AY) stars, perhaps the final stage of S Dor stars, are losing huge amounts of their mass as it is blown out into space.

Although eruptive variables, as a whole, do not receive the same level of attention from amateur astronomers that is typically directed toward pulsating variables, cataclysmic variables and eclipsing binaries, you will find many interesting stars claiming residence within this class of variable stars. With many of these stars, the challenge of observing variable stars possessing small amplitudes usually provides a reward of short periods. Of course, small amplitudes require the use of photometric methods, such as CCD and PEP, but the reward is that in some cases it's possible to observe one, or more, complete cycles in an evening. Amateur interest is growing with respect to several groups of stars found within this classification.

The B[e] stars were officially recognized in 1989 (IBVS 3323) to distinguish them from the related and similar γ Cassiopeia stars (GCAS). B[e] stars are almost universally understood to be spectral type O6–B9 stars, luminosity class III–V, that show variability with periods of much less than one day, normally with low amplitudes of a few percent. Instruments such as CCDs or photometers are usually considered necessary to observe these stars adequately.

The *FU Orionis* variables (FU Ori) are occasionally called *Orion variables*, *Orion population variables* or *nebula variables*, because many of them are connected in some way to nebulosity. FU Ori stars are T Tauri stars that are in a distinct stage of their evolution and outbursts are believed to be caused by instabilities within their accretion disk (see Chapter 5, "Cataclysmic Variables" for an explanation of accretion disks). Most of these stars are considered faint, even at their brightest but in many cases the flares, a sudden brightening, can be observed visually.

The γ Cassiopeiae stars have sometimes been called γ Eridani stars. γ Cas stars are rapidly rotating, luminosity class III and IV, B-type stars, possessing amplitudes reaching $1^m.5$ in V. B[e] stars are often called γ Cas stars if they are periodic. Comparable to B [e] stars, the observational methods used to observe γ Cas stars are similar but distinguishing between these two similar types of variables can be difficult and extra effort is obligatory when studying these stars.

The *irregular variable stars* exhibit a varied and interesting assortment of characteristics. Because the mixture of characteristics exhibited by these stars can become confusing very quickly, I've separated this large group of stars, within this book only, into three separate groups for easier study: irregular variable stars (I, IA and IB), the Orion variables (IN, INA, INB, INT and IN (YY)) and the rapid irregular variables (IS, ISA and ISB).

The group of *irregular variables* as a whole is generally poorly studied and contains many stars better cataloged into one of the other major groups of variable stars; however, insufficient information exists to do so. This subgroup of variable stars could become an excellent arena for amateurs to conduct real research and provide a tangible service since many of these stars are incorrectly classified. Rigorous, well-planned, and long-term observing programs may uncover interesting treasures hidden within this group of stars.

The *Orion variables* are connected in some way with nebulosity; they are usually embedded within a nebula or found near a nebula. Many are found in and around the Great Orion nebula and are so named. Other well-known nebulae that contain large numbers of these variables include: the Cone nebula (NGC 2264), the Flame nebula (NGC 2024), the Pelican nebula (IC 5070), and the Trifid nebula (M20).

The *rapid irregular variables* are similar to the Orion variables but have no apparent connection with nebulosity. Of growing interest to amateur astronomers, care must be taken when identifying rapid irregular variables as there is no strict boundary found between these stars and the Orion variables.

R Corona Borealis variables are a rare group of objects with perhaps only 30 true members known at this time. They remain close to maximum light for long periods of time but at unpredictable intervals they will undergo spectacular drops in brightness of up to 9^m in V. It may take one to three years before the star reaches maximum again. These stars are always popular and should be considered an excellent group of stars for study because of their dramatic fading, large amplitudes and relative brightness.

The *RS Canum Venaticorum* stars can be a confusing class of stars because this classification appears in the *GCVS* twice, within two of the main classes. First it appears as one type of eruptive variable stars that we will discuss here. This may be misleading because the mechanism for the variability is actually rotational modulation, with the surface brightness changing as a result of cool spots distributed unevenly across the surface of the star, yet RS CVn stars do not appear as one of the types of rotating variable stars. Second, the classification appears as a type of close binary eclipsing system according to the physical characteristics of the two stars. You will see this when we investigate eclipsing binary systems.

The *S Doradus* stars, sometimes called *Hubble–Sandage variables*, are members of a group of stars commonly designated as luminous blue variables (LBVs), though LBVs do not necessarily need to be blue, since the phenomenon is not restricted to early-type stars. These massive, luminous stars show dramatic mass ejection followed by periods of quiescence.

UV Ceti stars, also called *flare stars*, are late-type dwarf stars that undergo a sudden brightening at irregular time intervals. Theses flares are, in principle, the same kind of phenomenon as solar flares, but with much higher energies.

Wolf–Rayet (WR) stars are very luminous hot Population I stars[3] with temperatures between 30,000 and 50,000 K, notorious for their high mass-loss rate, approximately 10^{-5} $M_\odot$ yr^{-1} (this mathematical shorthand means 0.00001 solar mass/per year). WR stars form an important evolutionary phase through which all massive stars above a certain limiting mass pass when going from the main sequence to the end of their lives.

The variable star type, official (or most recognized) designation and a short description of the eruptive variable stars is shown in Table 3.1.

---

[3]There is some confusion in nomenclature, since the name is also used for central stars of planetary nebulae (emission Population II stars).

Table 3.1. Eruptive variable stars arranged in alphabetical order by type

Variable star type	Designation (and subclasses)	
Be	**Be**	B-type emission stars
FU Orionis	**FU**	young, T Tauri like stars
γ Cassiopeia	**GCAS**	periodic B[e] stars (sometimes called γ Eri stars)
γ Eridanus		an older name used for γ Cassiopeia stars
Irregular variables	**I** (four subclasses)	
	IA	I variables with early-type spectra
	IB	I variables with middle to late-type spectra
	IN	Orion variable stars (four subclasses)
	INA	IN variables with early-type spectra
	INB	IN variables with middle/late-type spectra
	INT	T Tauri stars
	IN(YY)	IN variables that are accreting matter
	IS	rapid irregular (two subclasses)
	ISA	early-type spectra
	ISB	middle/late-type spectra
R Coronae Borealis	**RCB**	eruptive and pulsating variable stars
RS Canum Venaticorum	**RS**	close binaries with H and Ca II in emission
S Doradus	**SDOR**	luminous blue variables (LBV) also known as Hubble–Sandage variables
UV Ceti	**UV**	(two subclasses)
	UV	spectral type K V–M V stars, flaring on time scales of minutes
	UVN	flaring Orion-type variables of UV type
Wolf–Rayet	**WR**	very hot stars losing huge amounts of material

Be (B[e] stars)

Observation Key

 ★ Bright stars

 Small amplitudes

Short periods

CCD or PEP

– It becomes more and more clear that, although the majority of Be stars are photometrically variable, not all of them could be properly called GCAS variables. Quite a number of them show small-scale variations not necessarily related to shell events; in some cases the variations are quasi-periodic. By now we are not able to present an elaborated system of classification for Be variables, but we adopt a decision that in the cases when a Be variable cannot be readily described as a GCAS star we give simply Be for the type of variability. GCVS

B[e] stars have often been called γ Cas stars[4] or even γ Eri stars[5] although the γ Cas variable stars are a distinct group of stars recognized within the *GCVS* and must be considered as a separate group. Carefully

---

[4]Gamma Cassiopeia variables, named for the bright, B0IVe star, γ Cas.

[5]Gamma Eridanus variables, named for the bright, M1IIIb star, γ Eri.

distinguishing between these two similar groups of variable stars requires patience and care. The name γ Eri stars is not longer used.

Within the *GCVS*, you will find more than 220 B[e] stars as well as almost 50 listed as uncertain (i.e. B[e]:). A handful of these stars are being observed as part of several long-term projects and amateur participation is encouraged (e.g. Dr. John Percy of AAVSO).

The B[e] stars did not become an interesting group of variable stars until after the spectrograph was developed and became a tool used by astrophysicists. An interesting phenomenon that occurs within the spectra of these stars is that both emission and shell lines can disappear completely and when this occurs, *a B[e] star is indistinguishable from a normal B-type star*. For unknown reasons, perhaps several years later, emission and shell spectra may again form within the normal B absorption spectrum.

The long-term variability of these stars is not strictly periodic, though it may be cyclic, meaning that the variability repeats but not in a completely predictable manner. In order to understand the nature of the variations of B[e] stars, it will be necessary to patiently observe a large number of them over the course of many years. Long-term observing campaigns are necessary for disclosing the nature of long-term variations and instruments are recommended.

FU (FU Orionis[6] stars)

– These variables are characterized by gradual increases in brightness by about 6^m in several months, followed by either almost complete constancy at maximum that is sustained for long periods of time or slow decline by 1^m– 2^m. Spectral types at maximum are in the range Aea– Gpea. After an outburst, a gradual development of an emission spectrum is observed and the spectral type becomes later. These variables probably mark one of the evolutionary stages of T Tauri-type Orion variables (INT), as evidenced by an outburst on one member, V1057 Cyg, but its decline (2^m5 in 11 years) commenced immediately after maximum brightness was attained. All presently known FU Ori variables are coupled with reflecting cometary nebulae. **GCVS**

Observation Key	
★	Bright stars
	Large amplitudes
🌑	Long periods
👁	Visual

---

[6]Orion, the Hunter, usually understood to be the son of Poseidon and a mortal mother.

Figure 3.1. Digitized Sky Survey image of FU Ori.

FU Orionis stars are pre-main sequence (PMS) stars, newly formed out of interstellar matter, that have not yet reached a sufficiently high central temperature to ignite nuclear reactions within their core. As a result of their young age, during their formation stage, the PMS stars receive their energy by release of gravitational energy (Figure 3.1).

PMS stars are usually classified as eruptive variables and within the *GCVS* they are subdivided into many subgroups that include the FU Ori stars. However, this classification, based strictly upon morphological photo-metric properties (light curves), is ambiguous and of only limited value.

Within the *GCVS*, you will find only about 10 FU Ori stars and six are classed as uncertain (FU:). The prototype of this group, FU Orionis is an interesting star that displays a range in brightness from 16^m5 to 9^m6. At its brightest, this variable is well within the range of binoculars or a small telescope.

GCAS (γ Cassiopeiae[7] stars)

– These are rapidly rotating B III–IVe stars with mass outflow from their equatorial zones. The formation of

---

[7] Cassiopeia, a queen, was married to King Cepheus.

equatorial rings or disks is often accompanied by temporary fading. Light amplitudes may reach $1^m.5$ in V. GCVS

γ Cas was the first star in which the B[e] phenomenon, the appearance of one emission line at one epoch within the spectrum, was observed in 1866.[8] An emission line arises from hot gas in an equatorial ring, shell or cloud surrounding the star. The B[e] stars should not be confused with, or mistaken for, the γ Cas group of stars and so care must be taken when distinguishing between the two types. As you probably suspect, distinguishing between B[e] and γ Cas stars is not easy.

Essentially, the difference between B[e] stars and γ Cas stars is that light variability is attributed to a shell or disk surrounding the γ Cas stars (Figure 3.2), whereas for B[e] stars, light variations are not necessarily related to shell events and in some cases the variations are quasi-periodic. In γ Cas stars, this shell of ejected material produces a P Cygni feature that is the hallmark of the circumstellar envelope.

The P Cygni characteristic, a spectral line having a redward displaced emission component with one or

Observation Key

 Bright stars

Small amplitudes

Short periods

CCD or PEP

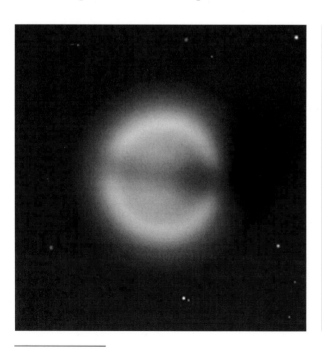

Figure 3.2. Artist's impression of a GCAS-type variable star showing the disk of material surrounding the star. Copyright: Gerry A. Good.

---

[8]By Pietro Angelo Secchi (1818–78), Italian Jesuit priest and astrophysicist.

more blueward displaced absorption components, is the result of one or more expanding gas envelopes surrounding the star and is named for the first star observed to display this particular line structure. The intensity of the P Cygni characteristic varies inversely with the rate of mass flow from the star. This distinction is beyond the abilities of most amateurs to detect, so a search of the literature will be necessary when compiling a list of observation targets. Be forewarned, no elaborated system of classification for B[e] variables exists and when a B[e] variable cannot be readily described as a γ Cas star, it is simply classed as a B[e] star. As a result, pay particular attention to the study of individual stars since their classification may change as new data is made available.

These variable stars can be interesting to observe but they do require a patient, long-term approach if you wish to extract any valuable information from them. As with B[e] stars, γ Cas stars are receiving increased interest from within the amateur community. There are almost 160 γ *Cas* stars found within the *GCVS*.

I (Irregular variable stars)

Observation Key

 Mixed stars
 Mixed amplitudes
 Mixed periods
 Visual, CCD/PEP

– Poorly studied irregular variables with unknown features of light variations and spectral types. This is a very inhomogeneous group of objects. IA (subtype) – Poorly studied irregular variables of early (O–A) spectral type. IB (subtype) - Poorly studied irregular variables of intermediate (F–G) to late (K–M) spectral type. GCVS

Within the *GCVS* you will find almost 1700 irregular variable stars and within this class of variable stars you will find a great mix of stars with a variety of spectral classifications, luminosity classes and physical characteristics. As stated within the official description, it consists of "a very inhomogeneous group of objects." For the amateur astronomer, this is an excellent environment in which to search for interesting stars. Most will require instruments but a good number are observable by visual means.

Generally, this is a *poorly studied* group of stars. Few astronomers have found the time to study these stars so you have an opportunity to make some interesting discoveries. However, because it is a poorly studied group of stars, you will experience some difficulty finding relevant research literature. Not much information exists regarding most of these stars and in many cases, the only information available will be that the star is identified within an *Information Bulletin* or *Name List*. Essentially, you'll be on your own but it can be appealing to the adventurous amateur seeking the challenges found away from the well-beaten path. If you decide to study this group of stars, check your work carefully and proceed with due concern for accuracy.

A good way in which to begin the study of one of these stars is to first determine its brightness. Make sure that you can actually see the star with your equipment because many of these stars are considered faint. Of the stars classified as "I," most are no brighter than magnitude 13! The good news is that over 20 get no fainter than magnitude 13, and so are well within the reach of many telescopes used by amateur astronomers. Seven of these stars have amplitudes equal to, or greater than one magnitude, meaning that they are observable by visual means. Of course, comparison stars may be difficult to find.

Second, take a look at the spectral type of the star in which you are interested. In some cases this isn't known but in many cases you can find it with a little research on your part. The spectral type may suggest some physical characteristic that will assist you in your study. For example, the variable star V398 Aur was identified as a variable star and listed as an irregular (I:) within *The 72nd Name List of Variable Stars*. The colon that follows the letter designation indicates the uncertain nature of this classification. V398 Aur is listed as an F0V (late, main sequence) spectral type. As you will discover, based upon its spectral type, this star falls within the Cepheid instability strip[9] found on the HR diagram. This star would certainly be a good candidate, at least in a statistical sense, for pulsational variability.

IN (Orion variable stars)

– Irregular, eruptive variables connected with bright or dark diffuse nebulae or observed in the regions of these

---

[9]See Chapter 4, "Pulsating Variable Stars," for a discussion of the Cepheid instability strip.

*nebulae. Some of these may show cyclic light variations caused by axial rotation. In the spectrum–luminosity diagram, they are found in the area of the main sequence and subgiants. They are probably young objects that, during the course of further evolution, will become light-constant stars on the zero-age main sequence (ZAMS). The range of brightness variations may reach several magnitudes. In the case of rapid light variations having been observed (up to 1^m in 1–10 days), the letter "S" is added to the symbol for the type (INS). This type may be divided into the following subtypes: **INA** (subtype) – Orion variables of early spectral types B–A or Ae. These are characterized by occasional abrupt Algol-like fadings. **INB** (subtype) – Orion variables of intermediate and late spectral types F–M or Fe–Me. F-type stars may show Algol-like fadings similar to those of INA stars; K–M stars may produce flares along with irregular light variations. **INT** (subtype) – Orion variables of the T Tauri type. Stars are assigned to this type on the basis of the following, purely spectroscopic, criteria: spectral types are in the range Fe–Me. The spectra of most typical stars resemble the spectrum of the solar chromosphere. The feature specific to the type is the presence of the fluorescent emission lines Fe II ll4046, 4132, emission lines [Si II] and [O I], as well as the absorption line Li I λ6707. These variables are usually observed only in diffuse nebulae. If it is not apparent that the star is associated with a nebula, the letter "N" in the symbol for the type may be omitted. **IN(YY)** (subtype) – Some Orion variables show the presence of absorption components on the redward sides of emission lines, indicating the infall of matter toward the stars' surfaces. In such cases, the symbol for the type may be accompanied by the symbols "YY."* GCVS

Of all the irregular variables, this group is probably the most visually enjoyable to observe because these stars are found within, or near, nebulae. Visiting a nebula in search of variable stars can be exciting by itself because nebulae are spectacular objects. Stopping to enjoy the sights along your journey in search of variable stars is something you shouldn't miss.

An observing approach similar to the other irregular stars is suggested; in other words, carefully watch for the irregular flares and be prepared to compare the variable with a pre-selected group of comparison stars.

Observation Key

★ Faint stars
 Small amplitudes
 Short periods
👁 Visual, CCD/PEP

IS (Rapid irregular variable stars)

*– Rapid irregular variables having no apparent connection with diffuse nebulae and showing light changes of about $0^m.5–1^m.0$ for several hours or days. There is no strict boundary between rapid irregular and Orion variables. If a rapid irregular star is observed in the region of a diffuse nebula, it is considered an Orion variable and is designated by the symbol INS. To attribute a variable to the type IS, it is necessary to take much care to be certain that its light changes are really not periodic. Quite a number of the stars assigned to this type in the GCVS 3rd Ed., turned out to be eclipsing binary systems, RR Lyrae variables, and even extragalactic BL Lac objects. **ISA** (subtype) – Rapid irregular variables of the early spectral types B–A or Ae. **INB** (subtype) – Rapid irregular variables of the intermediate and late spectral types F–M and Fe–Me. **GCVS***

Observation Key	
★	Faint stars
	Small amplitudes
	Short periods
	Visual, CCD/PEP

In general, another group of stars displaying duplicitous characteristics probably assembled by the Universe with the intent to challenge your observational skills. Within the GCVS, you will find approximately 230 rapid irregular variables including early (ISA), and intermediate through late spectral types (ISB) (see Figure 3.3).

Most of these stars are considered faint but are within the range of medium-to-large telescopes. A small number are within the reach of binoculars or small telescopes. As with the group as a whole, little information will be found on individual stars but rather than viewing this as an obstacle, you should

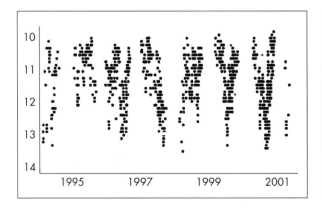

Figure 3.3. Light curve of the INSA-type variable star, RR Tau. Data provided by VSNET. Used with permission.

consider this an excellent opportunity to conduct some original research, perhaps resulting in ill-defined stars being better defined and properly placed within the classification scheme of the *GCVS*.

For example, CV Dra was discovered to be a short-period variable in 1960 but no further details were given other than that it was variable. The star was further investigated in 1961 but no confirmation of periodicity was evident so the star was classified as a rapid irregular variable. It is listed as one of the brighter rapid irregulars within the *GCVS* 4th Ed. However, in 1988, after further study, the star was determined to probably be an eclipsing binary of W UMa type.

RCB (R Coronae Borealis[10] stars)

Observation Key

 Bright stars
 Large amplitudes
 Long periods
Visual

– These are hydrogen-poor, carbon- and helium-rich, high-luminosity stars belonging to the spectral types Bpe–R, which are simultaneously eruptive and pulsating variables. They show slow non-periodic fadings by 1^m– 9^m in V lasting from a month or more to several hundred days. These changes are superposed on cyclic pulsations with amplitudes up to several tenths of a magnitude and periods in the range of 30^d–100^d. GCVS

Variables of the R Coronae Borealis type are hydrogen-poor, carbon- and helium-rich, high-luminosity stars belonging to the spectral types Bpe–R and are distinguished from other hydrogen-deficient objects by their spectacular dust-formation episodes. These stars are simultaneously eruptive and pulsating variables and exhibit some of the most spectacular behavior of any variable star.

They are apparently of low mass yet they possess high luminosity and at irregular intervals are known to manufacture thick dust clouds that can completely obscure the photosphere of the star. These changes are superposed on cyclic pulsations with amplitudes up to several tenths of a magnitude and with periods in the range 30–100 days (Figure 3.4).

The spectral class "R" is introduced here so a short explanation is needed. Within the Harvard spectral system, these stars are known as *carbon stars*. Carbon stars are late giant stars with strong bands of carbon

---

[10]The Northern Crown, Corona Borealis, represents a crown give by Dionysus to Ariadne, the daughter of Minos of Crete.

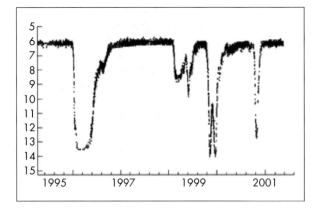

Figure 3.4. Light curve of the prototype RCB-type variable star, R Crb. Data provided by VSNET. Used with permission.

compounds and no metallic oxide bands. There are actually two spectral types known as carbon stars: R-type, similar to G5–K0 stars and displaying strong carbon spectral bands; and N-type that displays even stronger carbon spectral bands. The major difference between R- and N-type stars is the presence of the elements carbon and oxygen. The R-type stars are divided into decimal subtypes, whereas the N stars are divided using subtypes a, b and c. Today, carbon stars are usually classified as C-type stars with R-type stars known as early (hotter) C-type and N-type stars known as late (cooler) C-type stars. The temperature sequence for carbon stars varies from approximately G4 through M4 and it is well known that many C stars are variable.

The RCB stars are thought to be the product of a final helium shell flash or the coalescence of a binary white-dwarf system. These stars are interesting and important, first because they represent a rare, or short-lived stage of stellar evolution, and second because these stars regularly produce large amounts of dust and so they serve as laboratories for the study of dust formation and evolution (Figure 3.5).

Hydrogen-deficient stars also include the extreme helium (EHe) stars and the hydrogen-deficient carbon (HdC) stars. These stars are all supergiants ranging from B- to G-type with very little hydrogen found within their atmospheres. The HdC stars are distinguished from RCB stars by the absence of large-scale variability and IR excesses. IR excess is a phenomenon in which long-wavelength energy, that is infrared radiation, is detected at a level greater than can be simply explained as being produced by the star in question. Usually, the excess infrared radiation is accounted for by understanding that

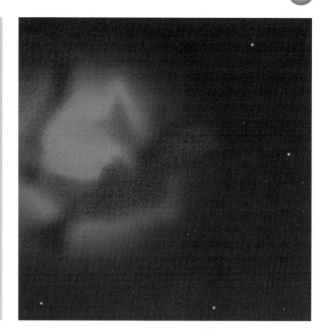

Figure 3.5. Artist's conception of an RCB-type variable star, illustrating the surrounding cloud of material responsible for obscuring the star's light. Copyright: Gerry A. Good.

short-wave energy is absorbed by some material and then re-emitted as a long wave. The EHe stars are hotter and with the exception of three RCB-like stars, do not show large-scale variability.

Despite their eye-catching light curves and spectacular behavior, the number of RCB variables is small. The *GCVS* lists about 40 with 14 *uncertain* but several other sources indicate as many as 45 RCB variables may be known.

RS (RS Canum Venaticorum[11] stars)

Observation Key

 Bright stars

Small amplitudes

Mixed periods

Visual, CCD/PEP

– This type is ascribed to close binary systems with spectra showing Ca II H and K in emission, their components having increased chromospheric activity that causes quasi-periodic light variability. The period of variation is close to the orbital one, and the variable amplitude is usually as great as $0^m.2$ in V. They are X-ray sources and rotating variables. RS CVn itself is also an eclipsing system. GCVS

---

[11] Canis Venatici, the Hunting Dogs, identified as Asterion and Chara, are usually depicted as greyhounds held on a leash by Boötes, the Herdsman.

As stated earlier, RS variables can be a confusing class of stars because this classification appears in the *General Catalog of Variable Stars* twice, in two of the main classes. As a group, these are relatively "new" stars since it wasn't until 1976 that Douglas Hall defined RS CVn binaries. RS CVn binaries have been, at least superficially, known as a subset of Algol-like eclipsing binaries, but with unusual properties that distinguish them from normal Algol-type variables.

Much speculation exists regarding the evolutionary status of these systems; in particular, regarding the number of stars with equal mass and because one star can be a red giant while the other is close to the main sequence. Such a configuration is difficult to understand. Several theories, some quite exotic, have been suggested to explain the formation of these interesting stars. They include: fission of a main sequence star, stars still in a state of pre-main sequence contraction, and evolved single stars with the more massive star losing some mass as it crosses the Hertzsprung gap.

RS variables, as a result of the underlying physics, can confound casual observation so extreme care must be taken when observing these stars. These chromospherically active stars vary in brightness on a variety of different time-scales: some are literally periodic, some are not strictly periodic and some can only be described in terms of a long time-scale. Some chromospherically active binaries, not showing eclipses, have been classified as ellipsoidal variables (see Chapter 6, "Rotating Variable Stars"). The variable star BH CVn is a non-eclipsing chromospherically active binary system that varies only as a result of the reflection effect. The reflection effect is described in Chapter 7, "Close Binary Eclipsing Systems."

Because these stars have large "spots," the brightness varies as the star rotates. This variation is commonly called the "wave" and can appear in the light curve superimposed on additional variability that might be resulting from eclipses, ellipticity, or reflection. Careful study of these stars can produce excellent data, including remarkably accurate rotation periods. Complex light curves should be expected but short periods in some systems allow detailed study of some of these stars.

SDOR (S Doradus[12] stars)

Observation Key

⭐ Bright stars
📈 Large amplitudes
😐 Mixed periods
👁 Visual, CCD/PEP

– These are eruptive, high-luminosity Bpec–Fpec stars showing irregular, sometimes cyclic, light changes with amplitudes in the range 1^m–7^m in V. They belong to the brightest blue stars of their parent galaxies. As a rule, these stars are connected with diffuse nebulae and surrounded by expanding envelopes. GCVS

The S Dor stars, occasionally called Hubble–Sandage stars and luminous blue variables, are a relatively new entry to the *General Catalog of Variable Stars*. They were officially recognized on 31 March 2000, when described within IBVS 4870, *The 75th Name List of Variable Stars* although their eruptions have been known, and observed, for centuries. Probably the best known S Doradus star is η (Eta) Carinae, visible from the southern hemisphere.

Between the years 1600 and 1800, astronomers occasionally noticed η Carinae as either a second- or fourth-magnitude star. It apparently varied in brightness, fluctuating between these two limits, until the late 1830s, when it became one of the brightest stars in the sky, remaining so for nearly 20 years. John Herschel called it "fitfully variable" and described its "sudden flashes and relapses" as it varied in brightness between magnitude +1 and –1. η Carinae eventually faded to eighth magnitude as the eruption ended and a circumstellar dust cloud formed. Today, η Carina teeters on the edge of catastrophic collapse and is watched carefully by astronomers from around the world.

UV (UV Ceti stars)

Observation Key

⭐ Faint stars
 Mixed amplitudes
 Fast outbursts
👁 Visual, CCD/PEP

– These are K Ve–M Ve stars sometimes displaying flare activity with amplitudes of from several tenths of a magnitude up to 6^m in V. The amplitude is considerably greater in the ultraviolet spectral region. Maximum light is attained in several seconds or dozens of seconds after the beginning of a flare; the star returns to its normal brightness in several minutes or dozens of minutes. UVN (subtype) – Flaring Orion variables of spectral types Ke–Me. These are phenomenologically almost identical to

---

[12]Dorado, the Swordfish, was formed by Johann Bayer in 1603 and appears in his *Uranometria* (star catalog).

UV Cet variables observed in the solar neighborhood. In addition to being related to nebulae, they are normally characterized by being of earlier spectral type and greater luminosity, with slower development of flares. They are possibly a specific subgroup of INB variables with irregular variations superimposed by flares. GCVS

UV Ceti variables, also known as flare stars, are late-type dwarf stars that undergo a sudden brightening at irregular time intervals. Their spectral type is K or M, but most are M[e] stars, that is they show emission lines in their spectrum. During a flare, the increase in the brightness of the star can be more than 6^m0. Interestingly, the amplitude of the flares increases with decreasing wavelength, that is, it is stronger in the U band than in the V band.

The time intervals between consecutive flares can be very different but they are usually between several hours and several days. Of course, exceptions occur. These flares are, in principle, the same kind of phenomenon as solar flares, but with much higher energies involved (Figure 3.6).

Observers making observation of UV Ceti stars should be prepared to conduct lengthy observations with the intent of catching a short flare or perhaps, several flares, over the course of an evening. Since the

Figure 3.6. Artist's conception of a UV Ceti-type variable star showing a bright flare erupting on the surface. Copyright: Gerry A. Good.

flares can reach maximum within several seconds, vigilance and patience will be required of the observer. Precise timing as well as precise determination of the brightness of the flare will be expected if your data is to have value.

WR (Wolf–Rayet stars)

– Stars with broadband emission features of He I and He II as well as C II-C IV, O II-OIV, and N II-N V. They display irregular light changes with amplitudes up to $0^m.1$ in V, which are probably caused by physical processes, in particular, by non-stable mass outflow from their atmosphere. GCVS

Wolf–Rayet (WR) stars, named after the French astronomers Charles Wolf and Georges Rayet, who discovered them in 1867, are as strange as the luminous blue variables. WR stars are luminous hot supergiants with temperatures comparable to those of normal O stars. However, they cannot actually be placed within this spectral class because of their eccentric spectra that display only emission lines and with little or no evidence of the most common of elements, hydrogen.

WR stars display luminosities that range between about 100,000 and a million times that of the Sun, at the limit or close to those of the LBVs. Considered rare – there are probably only a thousand or so within the Galaxy – they are at least more common than the LBVs. γ^2 Velorum, one of the sky's brighter stars, shining at $1^m.8$, is a double star comprising an O-type giant and a Wolf–Rayet star. Also like the LBVs, the Wolf–Rayet stars are losing mass at high rates; a ten-thousandth to a hundred-thousandth or so of a solar mass per year and the dominant element is not hydrogen, but helium

Presently, there are approximately 20 Wolf–Rayet stars found within the GCVS. Two types exist: nitrogen-rich (WN) and carbon-rich (WC). WN stars do contain small amounts of hydrogen, although the normal ratio of hydrogen to helium is reversed. In normal stars, there is about 10 times as much hydrogen as helium, whereas in WNs, there is typically 3–10 times as much helium as hydrogen. While the elements carbon and oxygen are effectively absent, WNs contain up to 10 times as much nitrogen relative to helium, and much more relative to hydrogen.

Chapter 4

Pulsating Variable Stars

Pulsating variables are stars showing periodic expansion and contraction of their surface layers. The pulsations may be radial or non-radial. A radially pulsating star remains spherical in shape, while in the case of non-radial pulsations the star's shape periodically deviates from a sphere, and even neighboring zones of its surface may have opposite pulsation phases.

GCVS

Of all the variable stars, pulsating stars, especially the Mira (M) and semiregular (SR) stars, are probably the most observed by amateur astronomers. This claim is easily understood when you consider that well over 22,000 pulsating variable stars are cataloged within the *GCVS* and several million pulsating stars probably exist within the Milky Way. By any measure, this is quite a selection and would keep you busy for many lifetimes. As well as providing a large number from which to choose, many pulsating stars have large amplitudes making them an excellent choice for visual observation.

Keeping in mind that only several million pulsating stars are believed to exist among several hundred billion stars within our Galaxy, pulsation is believed to be a brief phenomenon for most stars. Apparently, because it is a relatively brief occurrence, stellar pulsation is an unusual phenomenon and is therefore, as with all rare objects, fascinating. Upon closer examination, the locations of many of the pulsating variables within the HR diagram are certainly interesting.

As explained earlier, the majority of stars spend a large percentage of their lives on the main sequence. However, the majority of pulsating stars occupy a narrow, nearly vertical *instability strip* on the right-hand side of the HR diagram. As stars located within the main sequence begin to evolve and move off the main sequence to the right, some will eventually enter this instability strip and begin to pulsate. Pulsation will also cease when they exit this instability strip. Remember, this is not the only instability strip found within the HR diagram; it just happens to be the largest. We will look at other instability strips later when we discuss the various types of pulsating variable stars. Of course, stellar evolutionary time-scales are too long to observe the beginning and end of a single star's pulsation; however, a few stars have been observed in the final phase of this interesting evolutionary stage.[1]

There are several types of pulsation. One type is known as *radial pulsation*, and results when the star pulses, as if breathing, expanding and contracting equally in all directions. Another type of pulsation, known as *non-radial pulsation*, will be discussed in a few moments.

Sound waves resonating within a star's interior cause pulsation. You can approximately estimate the pulsation duration for these stars by determining the time required for a sound wave to cross the diameter of a model star. However, one of the sources of error when doing this is unrealistic model parameters. Usually, the model star is developed with a known radius and a constant density. We know that it's unrealistic to assume that a star has a constant density but approximations such as this are necessary and they are the working tools of astrophysicists. If you're looking for a pencil and a calculator, let me warn you that the ability to compute the innumerable dynamic conditions exhibiting continual changes, both subtle and gross, within a star is beyond the computational power of today's most advanced computers. However, it can be shown that the pulsation period of a star is inversely proportional to the square root of its *mean density*. The *period-mean density relation* explains the decrease in the pulsation period when you move down the instability strip from the tenuous supergiant stars to

---

[1] Polaris (α UMi), a classical Cepheid variable, has shown a sharp reduction in its oscillation during recent years.

the very dense white dwarf stars.[2] Think about this for a moment. Does sound travel faster in the atmosphere or, for example, underwater? Which is more dense? Of course, sound travels faster in the more dense water just as sound waves travel faster within the more dense stars. Fast traveling sound waves mean shorter periods.

Most of the classical Cepheids and W Virginis stars pulsate in what is called the *fundamental mode*. The long-period variables (LPVs) are probably also fundamental mode pulsating stars. In the fundamental mode, the gases of the star move in the same direction, inward or outward, at every point within the star. However, the RR Lyrae variables pulsate in either the fundamental or *first overtone modes*, with a few oscillating simultaneously in both. In the case of the first overtone, there is a single node between the center of the star and its surface defining the point at which gases move in opposite directions on either side of the node. The actual pulsation mode of the LPVs is the subject of considerable debate.

As briefly explained in the first chapter, when the core of a star is compressed, its temperature rises and thermonuclear energy increases. This energy-producing process, called the ε *(epsilon)-mechanism*, is found deep within the core of a star. Perhaps this is where pulsation originates?

But no, this is not the case, as we'll see. When first studied, pulsation was believed to originate within the various layers of the star, located above the core, and so the ε-mechanism was not considered to be the source of pulsation. Instead, Sir Arthur Eddington suggested that pulsating stars are thermodynamic heat engines. He believed that the stellar gases of the various layers within pulsating stars do work as they expand and contract. Accordingly, the net work performed by each layer during one cycle is the difference between the heat flowing into the gas and the heat leaving the gas. For maximum efficiency, the heat must enter the gas layer during the hottest part of the cycle and leave during the coolest part. In other words, the driving layers of a pulsating star must absorb heat at about the time of their maximum compression.

After much thought, Eddington suggested an unusual solution involving what he called a *valve mechanism*.

---

[2]Pulsating white dwarfs exhibit non-radial oscillations, and their periods are longer than predicted by the period-mean density relation.

He believed that if a layer deep within the star became more opaque upon compression, it could "dam up" the energy flowing toward the surface and as a result, this energy would push the upper layers of the star outward. As these layers were pushed outward, they became more transparent, allowing the trapped heat to escape. Once the heat escaped, the layer would fall back down to begin the cycle anew. In Eddington's own words, "To apply this method we must make the star more heat-tight when compressed than when expanded: in other words, the opacity must increase with compression."

Eddington's valve mechanism can successfully operate only within special layers of the star where the gases are partially ionized. In these *partial ionization zones*, part of the work done on the gases as they are compressed produces further ionization instead of raising the temperature of the gas. Ionization increases opacity, not temperature! The out-flowing energy is trapped by the ionization zones and the density of the zones increase.

As these layers are pushed outward by the increased pressure, their density begins to decrease and the ions begin to recombine and release energy. The temperature doesn't decrease much because the ions are recombining to release energy. The important thing is that the opacity of the layers decreases, with declining density during expansion. These layers of the star absorb heat during compression, are then pushed outward to release the heat during expansion, then fall back down again to begin another cycle. Astronomers refer to this opacity mechanism as the κ (*kappa-*) *mechanism*.

In some cases, the surfaces layers of stars do not move uniformly in and out. Instead, they display a more complicated type of *non-radial pulsation* in which some regions of the star's surface expand while other areas contract. In this situation, non-radial pulsation, the sound waves can propagate not only radially but also horizontally and produce waves that travel *around* the star. These non-radial oscillations are called *p-mode* oscillations because pressure is responsible for their compression and expansion.

For stars displaying non-radial pulsation, stellar material is not just moving in and then out, as if the star was breathing, but it is being sloshed back and forth. This sloshing of stellar gases produces another class of non-radial oscillations known as *g-modes*. The *g-modes* are produced by internal gravity waves. Because this sloshing cannot occur within stars displaying purely radial motion, there is no radial analog for the *g-modes* so they are found only within non-radial pulsating stars.

As you are beginning to understand, pulsation is complex and interesting and much can be learned from the study of pulsating stars. For example, the *g-modes* just mentioned involve the movement of stellar material deep within the star, while *p-modes* are confined near the stellar surface. Because they originate deep within the star, *g-modes* afford astronomers a probe into the very heart of a star. Just as important, the *p-modes* allow an inspection of the turbulent conditions within a star's surface layers. Of course, pulsation will tell much more than we've discussed here but this is just a beginning. With a basic understanding of pulsation, let's examine pulsating variable stars.

53 Persei stars are hot O- and B-type stars, usually classified as non-radial pulsating, with periods of less than a day. They are not officially recognized within the GCVS.

α Cygni stars are spectral type O through F, pulsating supergiant stars. These stars occasionally display variability that is typical of other classes of variable stars and as a result can be confused with other types.

The *β Cephei* stars have in the past been called *β Canis Majoris* stars and are normal early B-type giant stars with short periods lasting between two and seven hours. These stars occupy the *β Cephei instability strip*, located in a small area in the upper right-hand region of the HR diagram.

BL Boo stars are also known as *anomalous Cepheids* because they do not obey the period–luminosity relation of either the classical Cepheids or the Cepheids found within globular clusters.

A variety of designations have been used for Cepheids in the past and you must be careful when consulting old literature. You'll find the luminosity class IbII Cepheids of the characteristic spectral types F, G, or K; and the Type II Cepheids, known as *W Virginis stars* and in the past known as RRd stars. They are usually recognized as low–mass analogs to the classical Cepheids. Finally, you'll discover that the *δ Cepheids* are also known as classical Cepheids or Type I Cepheids.

As a group, Cepheid variables are interesting variable stars but you must be attentive during your study of these stars. The light curve of a pulsating star will depend upon its mass, structure and chemical composition. When attempting to classify Cepheids, it can be difficult to distinguish between Type I (classical) Cepheids, anomalous (BL Boo) Cepheids, and Type II (W Vir) Cepheids. Patience and attention to detail during your observations will help you distinguish between these groups of stars.

δ Scuti stars are low–amplitude, fast–pulsating stars of spectral type A to early F. These stars possess periods ranging from 30 minutes to eight hours. *δ* Scuti stars have been known to pulsate with up to 23 different modes.

γ Doradus stars are early F-type dwarf stars with short periods ranging from several tenths of a day to slightly more than a day. They became an official type within the *GCVS* in March 2000.

Slow irregular variables are generally poorly studied stars. Many of these stars are believed to be incorrectly classified and may actually be semiregular type variables. You will find two subtypes of slow irregular stars based on luminosity type (*giants and supergiants*).

λ Boötes stars are not listed within the *GCVS* but they are recognized as being metal-poor, Population I, A-type pulsating stars.

The *Mira* stars are a very popular group of variable star for amateur astronomers. They are also known as *long period variables* (LPVs) and are recognized as being late-type giants. Mira variables are known for large amplitudes, as much as 11 magnitudes.

Maia stars are predicted to exist but no true member of this group has ever been observed.

The *mid-B variable* stars were introduced in 1985 and are generally recognized to be B3–B8, luminosity class III–V stars with periods of approximately one to three days and amplitudes in the range of a few hundredths of a magnitude.

PV Telescopii stars are helium supergiants with periods less than a day and amplitudes in the range of a tenth of a magnitude.

RPHS variables, a new type of variable star that was officially recognized in 2001, are hot pulsating subdwarf stars previously known as EC 14026 stars.

RR Lyrae stars are fast radially pulsating A–F stars with amplitudes ranging from $0^{m}.2$ to 2^{m}, another group of stars well studied by amateur astronomers. There are three subtypes of RR Lyrae stars identified by the shapes of their light curves.

RV Tauri stars are radially pulsating supergiants,

spectral class F–M, characterized by light curves having double waves with alternating primary and secondary minima. There are two subtypes of RV Tauri stars identified by their mean magnitude.

The *semiregular* (SR) *variables* are similar to the Mira variables differing in that they generally have smaller amplitudes and shorter periods. Exceptions occur and some SR variables have long periods and large amplitudes. There are five subtypes of SR variables distinguished by spectral type and luminosity class demonstrating different amplitudes, periodicity and luminosity classes.

SX Phoenicis stars resemble δ Scuti variables but they display greater amplitudes.

UU Herculis stars are not recognized within the *GCVS* but are a possible transition phase between the asymptotic giant branch and the white dwarf evolutionary phases. UU Her, the prototype for this group of stars, seems to switch pulsation between fundamental and first overtone modes.

ZZ Ceti stars are non-radial pulsating white dwarfs with very fast periods and amplitudes reaching $0^m.2$. There are three subtypes of ZZ Ceti stars distinguished by their various spectra. The *GCVS* classifications are listed in Table 4.1.

53 PER (53 Persei stars)

– Non-radial g-mode pulsating stars of spectral type O9-B5 showing line profile variations with periods ranging from $0^d.16$ to $2^d.1$. **not recognized within the GCVS**

53 Persei[3] stars surround the instability zone of the β Cephei stars on the HR diagram with spectral types that range from O9 through B5. Suggestions have been made to add the 53 Persei star group to the variable stars observed in the range B3–B8 and called "mid B variables." Eventually, it was decided to separate the two groups and to classify the mid-B variables as "slowly pulsating B stars."

The concept of 53 Persei variables was introduced in 1979 for O8–B5 stars that show variability with periods

Observation Key

 Bright stars
 Small amplitudes
Short periods
CCD or PEP

---

[3] Perseus, ancient hero whose deed was to kill Medusa, a Gorgon, and eventually rescue the princess Andromeda from the sea monster Cetus.

Table 4.1. Pulsating variable stars arranged in alphabetical order by designation

Variable	Designation (and subclasses)		
53 Persei	*	**53 Per**	O9–B5 non-radial pulsating stars
α Cygni		**ACYG**	Be–Ae (emission) pulsating supergiants
β Cep		**BCEP**	classical β Cephei stars
		BCEPS	short-period β Cephei stars
BL Boo	*	**BLBOO**	anomalous Cepheids
Cepheids		**CEP**	radially pulsating F Ib–II stars
		CEP(B)	double mode pulsators
W Virginis		**CW** (two subclasses indicated below)		
		CWA	population II, period > 8^d
		CWB	population I, period < 8^d
δ Cepheids		**DCEP**	classical Cepheids, population I
		DCEP(S)	classical Cepheids, with overtone
δ Scuti		**DSCT**	A0–F5III/V pulsating stars
		DSCTC	low-amplitude δ Scuti stars
γ Doradus		**GDOR**	early-type F dwarfs showing multiple periods
Slow irregular variables		**L** (two subclasses indicated below)		
		LB	late-type giants
		LC	late-type supergiants
λ Bootis	*	**LBOO**	"p", non-magnetic, A–F, population I dwarfs
Mira		**M**	long-period late-type giants
Maia			stars predicted to exist but none have been found
mid B variables	*	**Mid B**	B3–B6 stars; periods = 1–3 days; amplitudes of up to $0^m.12$
PV Telescopii		**PVTEL**	helium supergiant Bp stars
RPHS		**RPHS**	Very rapidly pulsating hot subdwarf B stars (EC 14026 stars)
RR Lyrae		**RR** (three subclasses indicated below)		
		RR(B)	double-mode RR Lyrae stars
		RRAB	RR Lyrae stars with asymmetric light curves
		RRC	RR Lyrae stars with symmetric light curves
RV Tauri		**RV** (two subclasses indicated below)		
		RVa	radially pulsating supergiants with constant mean magnitude
		RVb	radially pulsating supergiants with variable mean magnitude
Semiregular variables		**SR** (five subclasses indicated below)		
		SRA	M, C, S, Me, Ce, Se giants/small amplitudes
		SRB	M, C, S, Me, Ce, Se giants/poorly defined periods
		SRC	M, C, S or Me, Ce, Se supergiants
		SRD	F, G, and K giants/supergiants
		SRS	semiregular pulsating red giants with short periods
SX Phoenicis		**SXPHE**	population II pulsating subdwarfs
UU Herculis	*	**UUHer**	high-latitude F supergiants
ZZ Ceti stars		**ZZ** (three subclasses indicated below)		
		ZZA	hydrogen pulsating white dwarfs
		ZZB	helium pulsating white dwarfs
		ZZO	showing He II and C IV absorption lines

* Indicates a designation found within the literature but not recognized within the *GCVS*.

Figure 4.1. Light curve of the prototype 53 Per-type variable star, 53 Per. Cycle phase is indicated along the horizontal axis. Data provided by the HIPPARCOS mission. Used with permission.

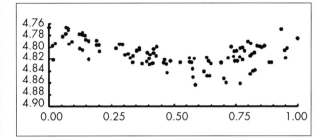

on the order of 24 hours. This is too long and too unstable to be associated with β Cephei type variability, and eventually the 53 Persei variable stars were classified as non-radial pulsating stars so it would be impossible to associate them with the radially pulsating β Cephei type variables (Figure 4.1).

Although 53 Per stars have long been suspected as non-radial pulsating stars, it is only very recently that pulsation instability within B-type stars is becoming understood. One of the characteristics of pulsation that recent studies are beginning to show is that pulsational instability is sensitive to fine details associated with metal opacities. Astronomers refer to all elements heavier than helium as metals.

Suggestions have also been made that the periodically variable B[e] stars may simply be 53 Per stars experiencing rapid rotation. However, this suggestion is usually considered to be wrong since 53 Per and B[e] stars are observationally distinct. Also, most B[e] variables have spectral types in the range which fall outside the instability domain for 53 Per stars.

This all suggests two different variability mechanisms for 53 Per and B[e] stars. In spite of many uncertainties, it seems very probable that the κ-mechanism, due to metal opacity, is responsible for driving a selective set of high-order, low-degree g-modes within 53 Persei stars. We have no idea why only modes within a narrow period range are excited and why rotation damps these modes. It is of great importance to determine the boundaries of the 53 Persei strip more accurately by searching for these stars in open clusters. The mere fact that B[e] stars are so common among low-metal systems, relative to systems with normal metal content, is impossible to explain if the driving mechanism is the same as in 53 Persei and β Cep stars, irrespective of anything else. In any case, there is ample evidence to justify the view that periodic variations are due to rotation and pulsation.

Because of their small amplitudes, these stars are best studied using photometric instruments such as CCDs or photometers.

ACYG (α Cygni stars)

– Non-radial pulsating supergiants of Bep–AepIa spectral types. The light changes with amplitudes of the order of $0^m.1$ often seem irregular, being caused by the superposition of many oscillations with close periods. Cycles from several days to several weeks are observed. GCVS

Observation Key

★ Bright stars
▨ Small amplitudes
⛰ Long periods
◉ CCD or PEP

α Cygni[4] variable stars are spectral type B and A, luminous, pulsating supergiant stars. You will notice that B- and A-type stars are located toward the upper left portion of the HR diagram where the relatively young, hot stars are positioned. The α Cyg classification now includes massive O- and late F-type stars since it has been determined that these stars also belong to the same stellar evolutionary sequence.

The star α Cyg (Deneb) is an A2Ib type star with an amplitude of $1^m.21$–$1^m.29$. As you can see, α Cyg is a hot supergiant itself (luminosity class Ib). This should not be surprising since this star is the model star (prototype) for all stars classified as α Cyg variables.

Because some α Cyg variables display variability that is typical for other classes of variable stars, they can be confused with other types of variable stars if you are not demanding when analyzing your data. The light curve of LT CMa is shown in Figure 4.2 to illustrate a representative α Cyg variable.

You can see that LT CMa produces a well-behaved light curve: a nice even climb and decline with no remarkable features. Beware: this behavior cannot be expected with all α Cyg stars. The light curve for Rho (ρ) Leo is also provided for comparison in Figure 4.3.

In contrast to these two light curves, the light curves of some α Cyg stars show dramatic features or random light fluctuations and as a result of this non-strict periodicity, meaning that light curve shapes vary from cycle to cycle, the periods are in fact "quasi" or

---

[4]Cygnus, the Swan, is an ancient constellation which appeared in Ptolemy's *Almagest* in the second century AD.

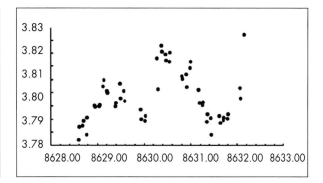

Figure 4.2. Light curve of the ACYG-type variable star, LT CMa. Julian dates are indicated along the horizontal axis. Data provided by the HIPPARCOS mission. Used with permission.

"pseudo-periods." Furthermore, in some cases, the variability of other types of variable stars mimics that of α Cyg type stars. One such example is the S Dor type star R71. While in its quiescent stage during 1983–85, it showed nearly the same type of optical oscillations as the normal α Cyg variables. It has even been suggested that the S Dor type stars be considered a subgroup of the α Cyg variables. To confuse the issue further, suggestions have been put forth that all S Dor type variables can be classified as P Cyg type stars since at maximum brightness all Balmer lines, some He lines and the lines of other ions show P Cyg profiles.

In recognition of this perplexity, it may help to remember that α Cyg variable stars are not: B[e] stars, β Cephei stars, low-mass 53 Per stars that do not pulsate in a regular manner, low-mass B-type super- and hypergiants that are probably post AGB proto-planetary nebulae stars, or low-mass F-type supergiants at high galactic latitudes (high galactic latitudes meaning located far away from the galactic plane), sometimes called UU Her stars.

In any case, these are interesting variables deserving close examination but they demand the utmost care

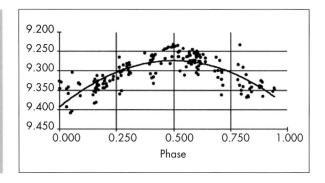

Figure 4.3. Light curve of the ACYG-type variable star, ρ Leo. Cycle phase is indicated along the horizontal axis. Data provided by the HIPPARCOS mission. Used with permission.

during their examination. As a result of their intrinsic brightness, suitable comparison stars can be difficult to find when observing these stars. Because of their small amplitude, α Cyg variables are probably best studied using a CCD or stellar photometer and are not considered good candidates for visual observation. The good news is that observing and studying these stars will strengthen your ability to detect subtle differences in complex data sets.

BCEP (β Cephei stars)

– Pulsating O8-B6 I–V stars with periods of light and radial-velocity variations in the range of $0\overset{d}{.}1$–$0\overset{d}{.}6$ and light amplitudes from $0\overset{m}{.}01$–$0\overset{m}{.}3$ in V. The light curves are similar in shape to average radial-velocity curves but lag in phase a quarter of the period, so that maximum brightness corresponds to maximum contraction, i.e. to minimum stellar radius. The majority of these stars probably shows radial pulsations, but some display non-radial pulsations; multiperiodicity is characteristic of these stars. BCEPS (subtype) A short-period group of β Cep variables. The spectral types are B2–B3 IVV; periods and light amplitudes are in the ranges $0\overset{d}{.}02$–$0\overset{d}{.}04$ and $0\overset{d}{.}015$–$0\overset{d}{.}025$ respectively, i.e. an order of magnitude smaller than the normally observed ones. GCVS

Observation Key	
★	Bright stars
	Small amplitudes
	Long periods
◉	CCD or PEP

β Cephei[5] variable stars, occasionally called β Canis Majoris stars, are a group of apparently normal early B giant and subgiant stars that display short-period light variations. The periods, lasting between two and seven hours, are too short to be explained by purely geometric effects, such as rotation and/or binary motion. Astronomers have recognized that the only remaining explanation requires stellar pulsation (Figure 4.4).

The interest in these variables for theoretical astrophysics was because theorists were not able to find a consistent explanation for the pulsational behavior of these stars and so the unknown driving mechanism for β Cep star pulsation remained one of the outstanding problems of stellar pulsation theory.

---

[5]Cepheus, the King of Joppa and one of the Argonauts who journeyed with Jason, was Cassiopeia's husband and Andromeda's father in ancient mythology.

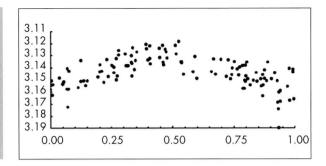

Figure 4.4. Light curve of the BCEP-type variable star, β Cep. Cycle phase is indicated along the horizontal axis. Data provided by the HIPPARCOS mission. Used with permission.

The variability of the radial velocity of β Cep was discovered at the beginning of the twentieth century at Yerkes Observatory. The star's period of variability was also determined to be 4^h34^m. In 1908 astronomers at Lick Observatory found that β CMa showed similar variation, and it turned out that this star became the first well-studied member of the group of β Cep variables. As a result, for several decades these stars were labeled β Canis Majoris stars.

The range of spectral and luminosity types among the β Cep stars force them into a small region within the HR diagram. This region is commonly labeled the *β Cephei instability strip*. However, it should be mentioned that some B[e] stars share this area in the HR diagram with the β Cep stars. Also, it should also be stressed that stars seen as β Cep stars at one time may later become B[e] stars. A most notorious example is β Cep itself, showing unprecedented strong emission in the core of the Hα_line in 1990, and conversely, one case is know of β Cep pulsation appearing in a well-observed B[e] star: 27 EW CMa developed a pulsation somewhere between 1987 and 1990.

Because of their small amplitudes, these stars are best studied using instruments; however, their relatively short periods allow one or more full cycles to be observed during an evening.

BLBOO (BL Boo stars)

The so-called "anomalous Cepheids," i.e. stars with periods characteristic of comparatively long-period RRAB variables, but considerably brighter by luminosity (BL Boo = NGC 5466 V19). **not recognized within the GCVS**

Astronomer's investigating variable stars have found within several dwarf spherical galaxies a group of Cepheid variables that do not obey the period–luminosity relation of either the classical Cepheids or the Cepheids found within the globular clusters, that is, the BL Her, W Vir, and RV Tau variables. These *anomalous Cepheids* have also been found in the Small Magellanic Cloud, but not as yet in the Large Cloud. The present interpretation of these stars is that they have masses of roughly $1.5M_\odot$ and are very metal poor.

In 1961 variable star 19 (hereafter V19) was discovered in NGC 5466. Star V19 has roughly the same color as the RR Lyrae variables in NGC 5466, but is also 1^m8 brighter. From this information it was inferred that V19 was either a W Virginis variable with a period of about 5 days or a foreground variable not associated with NGC 5466.

During this same time, based upon other observations, V19 was also suspected of being an eclipsing binary with a changing period, which led it to be included in *The General Catalog of Variable Stars (1969)* under the designation *BL Boötes*. This original observation was consistent with more recent ones, but the period and interpretation are both incorrect.

In 1972, V19 was observed once more and a new period of 0^d82 was determined. Based upon these new observations it was concluded that V19 was not a member of NGC 5466, but was instead a field RR Lyrae variable of Bailey type b. On the basis of these new observations V19 was included in the *Third Catalog of Variable Stars in Globular Clusters* where it was designated as number 19.

It was not until 1974 that is was suspected that V19 may be a member of NGC 5466 and that it was an unusual type of variable star. The reason for suspecting V19 to be a true member of NGC 5466 was that its radial velocity was consistent with membership in NGC 5466.

Observation Key	
	Bright stars
	Small amplitudes
	Long periods
◉	CCD or PEP

CEP (Cepheid stars)

– Radial pulsating, high luminosity (classes Ib–II) variables with periods in the range 1^d–135^d and amplitudes from several hundredths to 2^m in V (in the B band, the amplitudes are greater). Spectral type at maximum light is F; at minimum, the types are G–K and

the periods of light variation are longer the later the spectral type. The radial-velocity curves are practically reflections of the light curves, the maximum of the surface layer expansion velocity almost coinciding with the maximum light. CEP(B) (subtype) – Cepheids displaying the presence of two or more simultaneously operating pulsation modes (usually the fundamental tone with the period P_0 and the first overtone P_1). Their periods P_0 are in the range 2 to 7 days, with the ratio $P_1/P_0 \approx 0.71$. GCVS

Observation Key

 Bright stars

Mixed amplitudes

Long periods

CCD or PEP

At the beginning of 1784, only five variable stars, apart from novae and supernovae, were known. Four of these were what we now call long period variables (Mira-type variable stars) and one an eclipsing star, Algol. On September 10, 1784, Edward Piggot established the variability of η (Eta) Aquilae, while his friend John Goodricke showed that β Lyrae was variable. Shortly afterwards Goodricke found δ Cephei to vary.

While β Lyrae is the prototype of an important class of eclipsing variables, δ Cep and η Aql are what we now call *Cepheids* with periods of $5^{d}4$ and $7^{d}2$, respectively. Possessing visual light amplitudes of about $0^{m}9$, they are fairly representative of this class.

At one time, the term *Cepheid* meant any continuously varying star with a regular light curve and a period of less than approximately 35 days, unless it was known to be an eclipsing star. It is now recognized that the class defined in this way is heterogeneous, meaning that it contains stars in different mass ranges and evolutionary states. Stars with periods less than one day are now treated separately, mainly as RR Lyrae variables. Type II Cepheids and RV Tauri stars are also treated separately and their distinguishing features are dealt with later in this book. The remaining stars are called δ Cephei variables, Type I Cepheids, classical Cepheids, or simply and most frequently, Cepheids. You may find examples of Cepheid variables being subdivided into long-period, short-period, very short-period, ultra short-period and pseudo-Cepheids, but this terminology has not been generally adopted.

Cepheids are strictly periodic variables with periods ranging from about 1 day to about 50 days with a few extreme examples up to 200 days. The general form of the light curve varies smoothly as one moves from shorter to longer period stars. This is known as the *Hertzsprung progression*, named of course after the Danish astronomer who investigated it. The shorter-

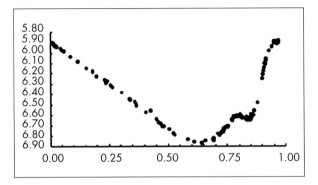

Figure 4.5. Light curve of the CEP-type variable star, X Cyg. Cycle phase is indicated along the horizontal axis. Data provided by the HIPPARCOS mission. Used with permission.

period variables have steep, narrow maxima. With increasing periods the relative widths of the maxima broaden. At periods of around 8–10 days the maxima often appear double. At shorter periods than this there are frequently bumps on the falling branch. The longer-period stars, 20–40 days, generally have very steep rising branches but the light curves of the longest-period Cepheids are more nearly sinusoidal (Figure 4.5).

The *McMaster Cepheid Photometry and Radial Velocity Data Archive* can be found at http://dogwood.physics.mcmaster.ca/Cepheid//HomePage.html. This Web site contains catalog data on the Type I (classical) and Type II (BL Her and W Vir) Cepheids, and extragalactic Cepheids. You will also find the *David Dunlap Observatory* database of galactic classical Cepheids.

CW (W Virginis[6] stars)

– These are pulsating variables of the galactic spherical component (old disk) population with periods of approximately $0^d.8$ to 35^d and amplitudes of $0^m.3$ to $1^m.2$ in V. They obey a period–luminosity relation different from that for δ Cep variables (DCEP). For an equal period value, the W Vir variables are fainter than the δ Cep stars by $0^m.7$–2^m. The light curves of W Vir variables for some period intervals differ from those of δ Cep variables for corresponding periods either by amplitudes or by the presence of humps on their

Observation Key

★ Bright stars
▦ Mixed amplitudes
☺ Mixed periods
◉ CCD or PEP

---

[6]Virgo, the Virgin. In the earliest records, she appears as a mother goddess, sometimes a wife to a creator or major god.

descending branches, sometimes turning in broad flat maxima. W Vir variables are present in globular clusters and at high galactic latitudes. They may be separated into the following subtypes: **CWA** *(subtype) – W Vir variables with periods longer than 8 days.* **CWB** *(subtype) – W Vir variables with periods shorter than 8 days.* **GCVS**

These pulsating stars are sometimes called *Type II Cepheids*. Care must be taken to differentiate these Cepheids from the δ Cep variables. The key is their periods of approximately $0\overset{d}{.}8$ to $35\overset{d}{.}0$ and amplitudes from $0\overset{m}{.}3$ to $1\overset{m}{.}2$ in V. Comparison of their light curves with δ Cep variables is interesting and with a little care taken during analysis, the difference can be detected.

When analysing light curves of W Vir variables, also look for the presence of humps on their descending branches, sometimes turning into broad, flat maxima. Since W Vir variables are present within globular clusters and at high galactic latitudes, you may be able to use this information to help further discriminate, or possibly, support your suspicion regarding these variables.

DCEP (δ Cephei stars)

Observation Key

 Bright stars
 Mixed amplitudes
Mixed periods
CCD or PEP

– These are the classical Cepheids, or δ Cephei variables. Comparatively young objects that have left the main sequence and evolved into the instability strip of the HR diagram, they obey the well-known Cepheid period-luminosity relation and belong to the young disk population. DCEP stars are present in open clusters. They display a certain relation between the shapes of their light curves and their periods. **DCEPS** *(subtype) – These are δ Cep variables having light amplitudes $< 0\overset{m}{.}5$ in V ($< 0\overset{m}{.}7$ in B) and almost symmetrical light curves (M–m approx. 0.4–0.5 periods); as a rule, their periods do not exceed 7 days. They are probably first-overtone pulsators and/or are in the first transition across the instability strip after leaving the main sequence.*

Traditionally, both δ Cep and W Vir stars are quite often called Cepheids because it is often impossible to discriminate between them on the basis of the light curves for periods in the range 3^d–10^d. However, these are distinct groups of entirely different objects in different evolutionary stages. One of the significant spectral differences between W Vir stars and Cepheids is

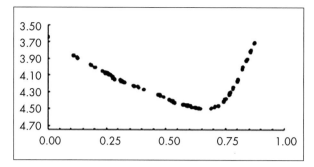

Figure 4.6. Light curve of the DCEP-type variable star, δ Cep. Cycle phase is indicated along the horizontal axis. Data provided by the HIPPARCOS mission. Used with permission.

the presence, during a certain phase interval, of hydrogen-line emission in the former and of Ca II H and K emission in the latter. **GCVS**

As indicated within the GCVS description, δ Cepheids and W Vir stars are difficult to discriminate between; in many cases, perhaps impossible without spectral analysis. As a result, you must take care when attempting to distinguish between these two types of similar variable stars (Figure 4.6).

DSCT (δ Scuti stars)

*– These are pulsating variables of spectral types A0–F5 IIIV displaying light amplitudes from $0^m\!.003$ to $0^m\!.9$ in V (usually several hundredths of a magnitude) and periods from $0^d\!.01$ to $0^d\!.2$. The shapes of the light curves, periods, and amplitudes usually vary greatly. Radial as well as non-radial pulsations are observed. The variability of some members of this type appears sporadically and sometimes completely ceases, this being a consequence of strong amplitude modulation with the lower value of the amplitude not exceeding $0^m\!.001$ in some cases. The maximum of the surface layer expansion does not lag behind the maximum light for more than 0.1 periods. DSCT stars are representatives of the galactic disk (flat component) and are phenomenologically close to the SX Phe variables. **DSCTC** (subtype) – Low amplitude group of δ Sct variables (light amplitude $< 0^m\!.1$ in V). The majority of this type's representatives are stars of luminosity class V; objects of this subtype generally are representative of the δ Sct variables in open clusters. GCVS*

Observation Key

★ Bright stars
▨ Small amplitudes
☺ Short periods
👁 CCD or PEP

δ Scuti[7] stars are spectral types A to early F pulsating variable stars, luminosity classes V to III. They pulsate in radial and non-radial pulsation modes, and possibly also gravitation modes, displaying periods between approximately 30 minutes and 8 hours. Photometric amplitudes are usually less than 1 magnitude. After the white dwarf ZZ Ceti variables, δ Scuti variables are the second most abundant type of pulsating variable star within our Galaxy.

The first mention of variability found within the star δ Scuti was made in 1900. At this time, a period was determined and δ Scuti was placed in the β Canis Majoris variable star group. Later investigations found that δ Scuti resembled the Cepheid variable stars rather than the hotter β Canis Majoris variables. In 1956, astrophysical developments pointed out the existence of a separate type of variable star.

Not surprisingly, the first δ Scuti stars discovered turned out to be unusual for their class because of their large photometric amplitudes. Only after 1965 could numerous discoveries of δ Scuti stars be made when photoelectric measurements with milli-magnitude precision became possible. As a result of the new precision available to astronomers, several systematic searches for δ Scuti variables were made in the late 1960s and early 1970s.

In 1970, after examining nine small-amplitude δ Scuti stars within the Hyades, astronomers proposed that all these variables with $P < 0^{\text{d}}.2$ should be called *ultrashort-period Cepheids*.

The very extensive photometric campaigns carried out 10–20 years later for specific δ Scuti stars and the discovery of dozens of stable frequencies show that the variability of δ Scuti stars is multiperiodic and regular in frequency.

A range of about $0^{\text{m}}.2$ is typical for the δ Scuti stars. The stars form a group which lies in an instability strip in the HR diagram which includes the classical Cepheids at its bright end and the pulsating white dwarfs at its faintest limit.

Today, the former dwarf Cepheids of the disk population are mostly called *high-amplitude δ Scuti stars*. Their pulsational behavior differs substantially from that of the stars with low amplitudes. They appear

---

[7]Scutum, or Sobieski's Shield, was a creation of Johannes Hevelius of Gdansk, who commemorated King John Sobeski III, who led the army which saved Vienna from the Turkish invasion of 1683.

to have only one or two radial modes excited with the remarkable exception of AI Velorum. In most cases, they oscillate in the fundamental mode or first overtone and closely resemble classical pulsational variables, like Cepheids or RR Lyrae stars. However, it is not clear whether or not non-radial pulsations are excited in a number of high-amplitude δ Scuti stars.

Since most known δ Scuti stars are brighter than 8th magnitude, small telescopes equipped with photometers are still used extensively for their study.

GDOR (γ Doradus stars)

– γ Doradus stars were officially added to the GCVS in March 2000 (The 75th Name-List of Variable Stars, IBVS 4870). They are described as early type F dwarfs showing (multiple) periods from several tenths of a day to slightly in excess of one day. Amplitudes usually do not exceed $0^{m}\!.1$. Presumably low degree g-mode non-radial pulsators.

Observation Key	
Bright stars	
Mixed amplitudes	
Mixed periods	
CCD or PEP	

δ Scuti stars, SX Phoenicis stars, and γ Doradus stars are three kinds of pulsating variable stars of similar spectral type and luminosity class. All three types are found near the main sequence in, or near, the Cepheid instability strip in the HR diagram.

δ Scuti stars comprise the largest class of these three types of stars and are subsequently the best studied. SX Phoenicis stars have similar periods but typically have much higher photometric amplitudes ($0^{m}\!.3$–$0^{m}\!.8$). γ Doradus stars are the most recently identified of the three types of variable stars just mentioned. They are typically early-F stars on, or just above, the main sequence in the HR diagram, and they are at or beyond the cool edge of the Cepheid instability strip. They exhibit photometric variability as large as $0^{m}\!.1$ in V on a time-scale of 0.5–3 days.

L (Slow irregular variable stars)

*–The light variations of these stars show no evidence of periodicity, or any periodicity present is very poorly defined and appears only occasionally. As for type L, stars are often attributed to this type because of insufficient study. Many type L variables are really semiregulars or other types. **LB** (subtype) – Slow irregular variables of late spectral type (K, M, C, S); as*

a rule, these are giants. This type is also ascribed in the GCVS to slow red irregular variables in the case of unknown spectral types and luminosities. LC (subtype) – Irregular variable supergiants of late spectral type having amplitudes of about 1 mag in V. **GCVS**

Irregular variable stars are slowly varying stars with no evidence of periodicity. Variable stars are frequently assigned to this class when their variability has been noted but not well-studied. This class of variable star is another excellent group deserving study by amateur astronomers.

One of the interesting aspects of studying these stars is that some semiregular (SR) variables experience periods of irregular variations so it is not entirely clear whether the irregular classification represents a fundamentally different type of variability.

As described within the *GCVS*, this group is composed of stars with spectral types K, M, C and S. The K-, M- and C-type have already been introduced (RCB stars). The S-type stars are presented here. An S-type star is a late-type giant, usually K5–M, that shows distinct bands of ZrO (zirconium oxide) in the blue and visual spectral regions. If these ZrO bands are weak or can be seen only at high levels of dispersion, the star is classed as MS so that a classification such as M4 S describes an M4 star with ZrO bands. Emission lines are seen within the spectra of those S-type stars that are variable. In general, S-type stars are rare; they are less abundant than C-type stars.

Observation Key	
	Mixed stars
	Mixed amplitudes
	Mixed periods
	Visual, CCD/PEP

LBOO (λ Boötis stars)

– The Lambda λ Boötis stars were defined within An Atlas of Stellar Spectra published in 1943 by Morgan, Keenan and Kellman. Apparently, λ Boö stars are a group of metal-poor, Population I, A-type stars. They occupy the same position on the HR diagram as Am stars that show a strong overabundance of many metals and the normal A-type stars. Not much is directly mentioned regarding variability as a characteristic of λ Boö stars.

When you think about these stars being A-type, surely a thought relating to the δ Scuti variables must briefly come to mind. In fact, some stars suspected of being λ Boö stars are listed as δ Scuti stars.

Observation Key	
	Bright stars
	Mixed amplitudes
	Mixed periods
	CCD or PEP

It is common practice to clearly identify the λ Boö stars using spectroscopic means. Specifically, this group of stars is identified by a deficiency in the iron-group elements (Sr, Fe, Ti and Sc), a deficiency of Mg and Ca compared to Fe, and they possess moderately large rotational velocities, among other characteristics. Subject to various interpretations, some of these methods of identification disqualify the λ Boö stars from being considered δ Scuti stars.

Certainly, λ Boö stars challenge our understanding of the various processes related to stars and they are fascinating members of the classical instability strip. Attempts to derive group properties with statistical methods have not been entirely successful because of the small number of unambiguously identified λ Boö stars. On the whole, classification appears to be obscured by incorrect group membership and it is therefore essential to provide a sufficiently large catalog of definitive group members before modeling the λ Boö phenomenon.

In the year 2000, an attempt was made to make a statistically sound analysis of their properties. In total, 708 program stars in the Galactic field, six open clusters, and the Orion OB1 association were observed. This resulted in 26 newly discovered λ Boötes stars and the confirmation of 18 candidates. Astroseismology was applied to the known members, resulting in 18 newly detected pulsators as well as 29 probably constant stars. The pulsational behavior of these stars is very similar to that of classical δ Scuti stars, so careful measures must be taken to detect the difference.

M (Mira stars)

– These are long period variable giants with characteristic late-type emission spectra (Me, Ce, Se) and light amplitudes from $2^m.5$ to 11^m in V. Their periodicity is well pronounced, and the periods lie in the range 80^d–1000^d. Infrared amplitudes are usually less pronounced than in the visible and may be $< 2^m.5$. If the amplitudes are $> 1^m$–$1^m.5$, but it is not certain that the true light amplitude is $> 2^m.5$, then the symbol "M" is followed by a colon, or the star is attributed to the semiregular class with a colon following the symbol for that type (SR).
GCVS

Observation Key	
★	Bright stars
	Large amplitudes
	Long periods
	Visual

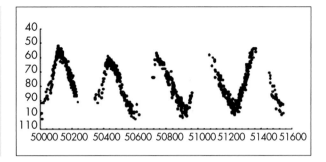

Figure 4.7. Light curve of the Mira-type variable star, R Leo. Julian dates are indicated along the horizontal axis. Data provided by VSNET. Used with permission.

Mira-type[8] variable stars, also known as long period variable (LPV) stars, are all similar (astronomers prefer to say "homogeneous") and probably the best studied of the pulsating red variables. The *GCVS* suggests three defining characteristics of the Mira variables: the spectral type is M[e], S[e] or C[e], the visual or photographic light amplitude must exceed $2^m.5$, and the period should be in the range 80–1000 days (Figure 4.7).

The spectral type indicates that Mira atmospheres contain strong molecular absorption features and are therefore cool because only within cool stars will molecules form. The outer atmospheres of these stars have temperatures below about 3800 K.[9] The atmospheres of these cool stars may be oxygen-rich (M[e]), carbon-rich (C[e]) or intermediate (S[e]). The emission lines, whose presence is signified by the "[e]," are an important characteristic of this type of variability as they are the signature of shock waves associated with pulsation. The amplitude boundary is essentially arbitrary and as a result a few stars that are physically similar to the Mira variables are classified as semi-regular (SR) type variables because their amplitudes fall short of $2^m.5$. The infrared light curves of the Mira variables, where most of the energy is emitted in these stars, and the total integrated luminosity have smaller amplitudes than those of visible light, although they are mostly over $0^m.5$. The large visual amplitudes arise from the combination that we are observing temperature variations from the blue side of the star's energy-distribution peak (the steepest portion of the visible energy window of the Planck curve at this temperature) and from changes in molecular-band strengths asso-

---

[8] Mira, the "wonderful," named by the Lutheran pastor and amateur astronomer named David Fabricius in 1595.
[9] When measured from absolute zero, temperature in degrees is properly called kelvins, not degrees Kelvin or Kelvin degrees.

ciated with these temperature changes. The very long periods tell us that Miras have very large radii. The upper limit to the period boundary is probably irrelevant. There are certainly stars with periods in the 1000–2000 day range that can be considered Mira variables.

The Mira variable stars are of great interest to astronomers, primarily because they represent a very short-lived phase in stellar evolution and in the HR diagram they are found at the very tip of the *asymptotic giant branch* (AGB), so their next evolutionary step is expected to be a rapid move across the diagram to become planetary nebulae. There are various studies that suggest the period of a Mira variable is a good indicator of the stellar population to which it belongs. Long-period variables with periods of around 200 days belong to the same, old population as do the metal-rich globular clusters. Longer-period Mira variables are more massive and/or more metal rich. Consistent with this picture, but contrary to popular belief, there is no evidence that Miras systematically evolve to longer periods as they age. Miras are also useful as distance indicators as they obey a period–luminosity relation that can be expressed either in terms of the total (*bolometric*) luminosity or in terms of the near-infrared magnitudes.

It remains uncertain whether Miras pulsate in the fundamental or first overtone modes. While there are theoretical reasons for favoring the fundamental mode, the observational evidence favors the overtone. Miras are rapidly losing mass, perhaps as much as 10^8–10^4 M$_\odot$ yr^1, although the mechanism for this is not well understood. The mass-loss rates are statistically correlated with the pulsation period, the bolometric light amplitude and the shape of the light curve. The most highly evolved Miras are surrounded by the material they have ejected, rendering them optically faint but strong infrared sources. The very long period Miras that have evolved from the most massive progenitors and have the most mass to lose, have particularly thick shells. Some of these circumstellar shells also produce SiO, H_2O, and/or OH maser emissions that are detectable at radio frequencies.

With careful examination, you will discover that the light curves for Mira variables are not identical from cycle to cycle and the brightening at maximum often varies by a magnitude or more from one cycle to the next. Period changes are also observed in some of these stars. Period change can be seen particularly clearly in R Aql and R Hya, because they may be undergoing helium shell flashes.

Within the *GCVS*, there are over 5,200 known Mira variables, with another 940 suspected. Clearly, there is a sufficient number of long period variables to study for many years to come.

MAIA (Maia stars)

– In 1955, the Maia[10] variable stars were predicted to exist within the spectral range B7–A2, possess periods between 1 and 4 hours and reside within the HR diagram between β Cephei and δ Scuti stars. Maia, the prototype for this group of stars, has a long history of suspected variability. Together with a second star, γ Ursa Minoris (A3 II–III or A3 V), these two stars have been suggested to form the ends of the hypothetical Maia sequence. Both stars are listed within the Yale Catalog of Bright Stars as variable.

Over the years, several searches have been conducted in an attempt to unambiguously locate the Maia variables. As recently as 1987 it was concluded that their existence was doubtful because no variability had been detected within the small group of suspects. Recently, a search using the Hipparcos epoch photometry database was conducted in an effort to search for the elusive Maia pulsating variable stars. Several hundred stars were considered, and several dozen stars studied in detail; however, only a handful are possible variables: three are possible shallow eclipsing variables; three have possible periods in the range $0^{d}.25$-$0^{d}.5$ but their amplitudes are so small that they are probably non-variable. γ UMi shows an irregular variation ranging from $0^{d}.0143009$ to $0^{d}.14335$ and a $\Delta m < 0^{m}.05$. It has also been suggested that γ UMi may be surrounded by a short-lived tenuous shell.

Maia and γ UMi are now regarded as photometrically constant. Perhaps like the abominable snowman, these stars may exist but are extremely difficult to find.

---

[10]Named for Maia (20 Tau = HR 1149 = HD 23408), one of the brighter stars found within the Pleiades star cluster.

mid-B (Middle B variable stars)

– In 1985 a class of "mid-B" variable stars was introduced within the literature. These stars are generally B3–B8 spectral type, luminosity class III–V. They have periods of 1^d–3^d and amplitudes of light variations of a few $0^m.01$. The color variations are in phase with the light variations, and the color-to-light ratio remains constant despite variability in amplitudes on a cycle-to-cycle and even a year-to-year base. **not recognized within the GCVS**

Observation Key	
★	Bright stars
▦	Small amplitudes
☺	Mixed periods
👁	CCD or PEP

In recent decades, astronomers have begun to realize that many, if not all, early-type stars show some kind of intrinsic variability the causes of which are unknown. A number of theoretical studies seem to indicate that the current equilibrium models of massive stars disagree with the observation to some degree. It is likely that our lack of understanding of the causes of the variations of the early-type stars is related to our poor understanding of their precise structure. The study of early-type variable stars is therefore not only of interest within the context of stellar pulsation theory but also when considered within the broader context of the theory of stellar structure.

Numerous observational studies of early-type variables have been carried out recently and the view that emerges is somewhat confusing. Some astronomers distinguish up to nine different classes of variable B stars. The distinctions between various classes seem to be based not only on a comparison of the characteristics of the variations but also on some *a priori* information. On the other hand, there is a temptation to disregard the accepted boundaries placed between most classes, and to call, for instance, a "β Cephei star" any B-type star that shows variability on a short time-scale compatible with a radial pulsation mode. A similar kind of reasoning is adopted by some when the descriptive term *slow variables* was introduced to name the very mixed group of stars that vary on a time-scale significantly longer than the fundamental period of radial pulsation.

These classification problems are partly caused by our inability to define the parameters that are relevant to describe the variability, and so to the very problem of our inability to isolate instability mechanisms that operate in these stars. Probably, it is, however, also true that at least some of the ambiguity in classifying the early-type variable stars is due to the diverse observational techniques that are used for detecting and describing these stars.

The distinction between these groups of stars is primarily based on the morphology of their light curves. The physical conditions in the stars of these few classes are different, and it is therefore probable that the variety in the morphologies of their light curves indicates that physically distinct mechanisms are responsible for their variations.

PVTEL (PV Telescopii stars)

Observation Key

 Bright stars
 Small amplitudes
 Short periods
👁 CCD or PEP

– These are helium supergiant Bp stars with weak hydrogen lines and enhanced lines of He and C. They pulsate with periods of $0\overset{d}{.}1$ to $1\overset{d}{.}0$, or vary in brightness with an amplitude $\approx 0\overset{m}{.}1$ in V during a time interval of about a year. GCVS

PV Telescopii stars in the past were called helium stars, helium horizontal branch stars, extreme helium (EHe) stars, and hydrogen-deficient binary (HdB) stars. Small amplitude light and radial velocity variations of the R Coronae Borealis (RCB) variable RY Sgr have long been identified with radial pulsation. Thereafter, small-amplitude light variations were observed in other RCB stars as well as in extreme helium (EHe) stars and in hydrogen-deficient binary (HdB) stars.

All three groups of stars are characterized by an extremely low surface abundance of hydrogen and low surface gravity. It has been known for some time that the behavior of the small-amplitude variations in some RCB stars is roughly periodic and as a result has been attributed to stellar pulsation. Evidence for periodic light variation of the HdB star KS Per was found in 1963 and the EHe star HD 160641 in 1975. In recent years more substantial evidence for periodic light variation of EHe and HdB stars has become available, beginning with the detection of a ~ 21 day period in the EHe BD +1° 4382 in 1985.

Although similar in many respects, RCB, EHe, and HdB stars represent quite distinct groups. EHe and RCB stars are apparently single stars with a surface abundance of C (carbon) and N (nitrogen). RCB stars are distinguished by the presence of a large infrared excess and by an occasional deep light minimum. HdB stars have low surface abundance of C and are all single-lined spectroscopic binaries.

As a simple consequence of their energy budget, the evolution of luminous blue stars invariably occurs on time-scales of a few hundreds to a few thousands of years. Extreme helium stars can be no exception. Their surfaces are predominately helium, with a few percent of carbon and nitrogen and in general a negligible contamination by hydrogen. They are almost certainly of low mass. The majority show small-amplitude variations on time-scales of 1–20 or more days, and should correctly be classified as PV Tel variables.

The principal question posed by the extreme helium stars is in regard to their evolutionary origin. Two principal hypotheses have emerged with the chief difference between the two being whether the progenitor is a single or a binary white dwarf. The general properties of both hypotheses are simply referred to as the late thermal pulse (LTP) model and the merged binary white dwarf (MBWD) model.

Being rare and unusual, several extreme helium stars were observed within the first two years of operation of the *International Ultraviolet Explorer* (IUE: 1978–79) and nearly all had been observed by the mid 1980s. Upon recognizing the great sensitivity of their fluxes to temperature, a series of second epoch observations of extreme helium stars was made with IUE in the early 1990s.

In the meantime it had been recognized that many extreme helium stars show photometric variability on time-scales of hours to weeks and the most luminous are likely to pulsate. Clearly, cyclic changes in the flux due to pulsation would be as easy to measure but could mask any secular changes due to evolution. Such cyclic changes would, however, be extremely useful for measuring the temperature and radius variations associated with the pulsation and could, in conjunction with radial velocity measurements, help to ascertain the radius of a pulsating helium star, independent of distance. Should the secular change be large compared with the cyclic change, then a measurement of the cyclic changes at one end of the secular vector would indicate the overall uncertainty in the length and direction of that vector.

When examining the observed properties of the hydrogen-deficient stars, it will be found that the periods of the variable hydrogen-deficient stars decrease as the effective temperature increases and the non-variable hydrogen-deficient stars tend to have small luminosity-to-mass ratios.

RPHS (Rapid pulsating hot subdwarf variable stars)

Observation Key

★ Bright stars
▨ Mixed amplitudes
▲ Mixed periods
👁 CCD or PEP

– The prototype star for this group of variable stars is EC 14026-2647. It appears to be an sdB star in a binary system that exhibits low-amplitude light variations, presumably as a result of stellar pulsation, with a main period of 144 s and a secondary period near 134 s, making it one of the shortest-period stellar pulsations known. EC 14026-2647 was discovered during the Edinburgh–Cape (EC) Blue Object Survey conducted to search for blue stellar objects. The survey is responsible for detecting many new hot subdwarfs, white dwarfs, cataclysmic variables, apparently normal B stars at high galactic latitudes, blue horizontal branch stars and bright quasi-stellar objects. Several new ZZ Ceti stars were reported along with a new member of the rare AM CVn class of hydrogen-deficient binaries.

Theses stars are now officially classified as very rapidly pulsating hot (subdwarf B) stars (RPHS) within the *General Catalog of Variable Stars* (IBVS 4870, 31 March 2000), however, because they are relatively new, they have not received much attention from amateur astronomers.

RR (RR Lyrae stars)

Observation Key

★ Faint stars
▨ Small amplitudes
 Short periods
 CCD or PEP

*Radially pulsating A–F stars having amplitudes from $0^m.2$ to 2^m in V. Cases of variable light-curve shapes as well as variable periods are known. If these changes are periodic, they are called the "Blazhko effect." Traditionally, RR Lyr stars are sometimes called short-period Cepheids or cluster-type variables. The majority of these stars belong to the spherical component of the Galaxy; they are present, sometimes in large numbers, in some globular clusters, where they are known as pulsating horizontal-branch stars. Like Cepheids, maximum expansion velocities of surface layers for these stars practically coincide with maximum light. **RR(B)** (subtype) – RR Lyrae variables showing two simultaneously operating pulsation modes, the fundamental tone with*

*the period P_0 and the first overtone P_1. The ratio P_1/P_0 ≈ 0.745. **RRAB** (subtype) – RR Lyrae variables with asymmetric light curves (steep ascending branches), periods from 0^d3 to 1^d2, and amplitudes from 0^m5 to 2^m in V. **RRC** (subtype) RR Lyrae variables with nearly asymmetric, sometimes sinusoidal, light curves, periods from 0^d2 to 0^d5, and amplitudes not greater than 0^m8 in V. **GCVS***

RR Lyrae stars have been studied for over 100 years. Their study initiated from the end of the nineteenth century when astronomers began an increasingly close scrutiny of globular clusters. It was during these observations that the first short-period variable stars were discovered. Although the first globular cluster variable to be discovered was a nova which erupted in M80 in the year 1860, it took another three decades before E.C. Pickering reported the discovery of a second globular cluster variable near the center of M3.

During the following few years a small number of bright globular cluster variables began to be discovered. In response to these discoveries, Solon I. Bailey began a program of globular cluster photography at the Harvard College Observatory station in Arequipa, Peru in 1893. Upon close examination of these photographic plates, Williamina Fleming discovered a variable star in the globular cluster w Centauri in August of that year. Following Fleming, Pickering detected six more variable stars in the same globular cluster in 1895 and this began the flood of ensuing discoveries. Between 1895 and 1898 more than 500 variable stars were discovered in globular clusters. These first short-period variables became known as the RR Lyrae variable stars and during the past century the number of RR Lyrae stars has grown to outnumber known members of any other well-defined class of variable star.

Of historical note, E.C. Pickering's 1889 discovery of a variable in M3 was probably a Cepheid rather than an RR Lyrae star but it does serve to demonstrate the initial confusion in distinguishing these two types of variable stars. As another example, Sir Arthur Eddington included RR Lyrae in a table of important Cepheid variables in his influential book *The Internal Constitution of the Stars*. Other astronomers, such as Henry Norris Russell, drew the distinction between Cepheids and RR Lyrae stars more precisely. Despite similarities with pulsating Cepheid variables the RR Lyrae variable stars from the first decades of this century were usually looked upon as a distinct class of variables, though having considerable kinship to the Cepheids.

Today, it is known that all RR Lyrae star are low-mass horizontal branch stars in the core helium burning stage of evolution and this has provided additional argument for distinguishing them from the higher-mass classical Cepheids. However, there have been adjustments from time to time in the types of variable star included in the class of RR Lyrae stars. A particular confusion arose with the short-period pulsating variable stars which are now usually called δ Scuti when they are classed as Population I stars, or SX Phoenicis stars when determined as Population II stars.

Referred to as cluster-type variables in the past for obvious reasons since they can occur in considerable numbers within globular clusters, you may occasionally find this classification still used today. RR Lyrae variable stars are giant radial pulsating stars with periods of $\sim0^{d}2$ to $1^{d}0$. Based on their light curves, the RR Lyrae stars can be subdivided into two main groups. The RRab stars have relatively large light amplitudes with visual amplitudes of about one magnitude being common. Their light curve is asymmetrical with a steep rising branch. The RRab Lyrae stars are believed to be pulsating in their fundamental mode and generally have periods of $\sim0^{d}4$ to $1^{d}0$. The RRc variables have smaller light amplitudes and their light curves are more nearly sinusoidal in shape. These variable stars are believed to be pulsating in their first overtone (sometimes called the first harmonic mode) and generally have periods of $\sim0^{d}2$ to $0^{d}5$.

Many of the RR Lyrae variables display long-term modulations of their light curves. This phenomenon is known as the *Blazhko effect*. Its cause is still not well understood. The modulation periods are generally in the range of 20–200 days and the effect on the light curve can be quite marked. For instance, in RR Lyrae itself the visual amplitude of the star varies by about $0^{m}3$ over the *Blazhko cycle* and the shape of the light curve changes. For some variables the *Blazhko effect* itself is modulated on an even longer time-scale. In the case of RR Lyrae the tertiary period is ~3.8–4.8 years.

The study of RR Lyrae stars has contributed to almost every branch of modern astronomy including: being used as tracers of the chemical and the dynamic properties of old stellar populations within our own and nearby galaxies; serving as standard candles, indicating the distances to globular clusters, to the center of the Galaxy, and to neighboring Local Group systems; and serving as test objects for theories of the evolution of low-mass stars and for theories of stellar pulsation.

RV (RV Tauri stars)

*– These are radially pulsating supergiants having spectral types F–G at maximum light and K–M at minimum. The light curves are characterized by the presence of double waves with alternating primary and secondary minima that can vary in depth so that primary minima may become secondary and vice versa. The complete light amplitude may reach 3^m–4^m in V. Periods between two adjacent primary minima (usually called formal periods) lie in the range 30^d–150^d (these are the periods appearing in the Catalog). Two subtypes, RVa and RVb,[11] are recognized: **RVa** (subtype) – RV Tauri variables that do not vary in mean magnitude. **RVb** (subtype) – RV Tauri variables that periodically vary in mean magnitude with periods from 600 to 1500 days and amplitudes up to 2^m in V. **GCVS***

	Observation Key

RV Tauri[12] stars appear to be *post-asymptotic giant branch*[13] (AGB) stars, evolving from the red giant to the white dwarf phase on a time-scale of thousands of years. They are spectral type F or G when brightest and G or early K when faintest. The visual light curves display double waves with alternating primary and secondary dips to minimum light that can vary in depth. The primary minimum brightness may become secondary minimum brightness.

The double-wave light curves probably result from a resonance between fundamental mode of pulsation and the first overtone mode. The amplitude in visual light is usually between one and two magnitudes although it may exceed three magnitudes. The periods between two adjacent primary minimum declines, usually called the formal period, is approximately 30–

---

[11] Within the *GCVS* the subtypes RVa and RVb are listed as RVA and RVB. This should be avoided because of the possible confusion with the spectroscopic subtypes (i.e. A, B, and C).

[12] Taurus, in Greek mythology, the snow-white bull which, carried Europa off, only to be revealed as Zeus in disguise.

[13] The second time that a star climbs the giant branch, it has a graphical track on the HR diagram that is crudely asymptotic to the first track and so is called the asymptotic giant branch.

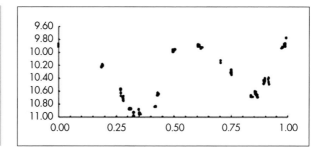

Figure 4.8. Light curve of the RV-type variable star, SZ Mon. Cycle phase is indicated along the horizontal axis. Data provided by the HIPPARCOS mission. Used with permission.

150 days. The cause of the slow variation in the mean amplitude, known as the *RVb phenomenon,* is not understood.

Light curves for RV Tauri stars are semiregular, and sizable variations are seen from one cycle to the next (Figure 4.8). The longer-period stars have a tendency to be less regular than the shorter-period stars. The phase of the U-B and B-V color curves precedes that of the visual light curve by up to a quarter of a period. Interchanges of deep and shallow minimum brightness occasionally occur and may be abrupt or gradual. Stretches of irregular or chaotic behavior have also been recorded.

The RV Tau stars are subdivided as a result of their long-term behavior. Those stars with clear long-term variability are classed as RVb and those without such variations as RVa. The RVb type are periodic with periods in the range of hundreds to thousands of days. The RV Tau stars can be oxygen- or carbon-rich and have also been subdivided into groups A, B, and C on the basis of their spectra.

The aging process of RV Tauri stars is equivocal. They have extended atmospheres, are undergoing mass loss, and TiO absorption bands are occasionally observed in the optical spectra near minimum light which is also the coolest phase. The TiO absorption bands indicate a later spectral type, perhaps beginning at M2. Dust-shells, as evidenced by their strong infrared emissions, surround some RV Tau stars. They may be AGB stars executing blue loops within the HR diagram following a helium-shell flash, or they could be post-AGB stars in the process of losing the last remnant of their atmospheres as they turn into white dwarfs.

RV Tauri stars are closely related to type II Cepheids, also found in metal-deficient globular clusters, and they occupy the same instability strip in the HR diagram as lower luminosity and shorter-period stars. They also have similarities to the semiregular variables, in particular SRd and the UU Her groups.

Additional examples of RV Tauri variable stars are DF Cyg, AC Her, U Mon, R Sct and RV Tau. R Scuti is classed as an RVa because of its period of 146.5 days.

The observed vs calculated (O-C diagrams) light curves of RV Tauri stars are known to be dominated by the effects of random cycle-to-cycle fluctuations in the period, as is the case in Mira stars.

SR (Semiregular variable stars)

*– Giant or supergiants of intermediate and late spectral types showing noticeable periodicity in their light changes, accompanied or sometimes interrupted by various irregularities. Periods lie in the range from 20 to >2000 days, while the shapes of the light curves are rather different and variable, and the amplitudes may be from several hundredths to several magnitudes (usually 1^m-2^m in V). **SRA** (subtype) – Semiregular late-type (M, C, S or Me, Ce, Se) giants displaying persistent periodicity and usually small (< $2^m.5$ in V) light amplitudes (Z Aqr). Amplitudes and light-curve shapes generally vary and periods are in the range of 35–1200 days. Many of these stars differ from Miras only by showing smaller light amplitudes. **SRB** (subtype) – Semiregular late-type (M, C, S or Me, Ce, Se) giants with poorly defined periodicity (mean cycles in the range of 20–2300 days) or with alternating intervals of periodic and slow irregular changes, and even with light constancy intervals (RR CrB, AF Cyg). Every star of this type may usually be assigned a certain mean period (cycle), which is the value given in the Catalog. In a number of cases, the simultaneous presence of two or more periods of light variation is observed. **SRC** (subtype) – Semiregular late-type (M, C, S or Me, Ce, Se) supergiants (Mu Cep) with amplitudes of about 1 mag and periods of light variation from 30 days to several thousand days. **SRD** (subgroup) – Semiregular variable giants and supergiants of F, G, or K spectral types, sometimes with emission lines in their spectra. Amplitudes of light variation are in the range from 0.1 to 4 mag, and the range of periods is from 30 to 1100 days (SX Her, SV UMa). **SRS** (subgroup) – Pulsating red giants with short periods (several days to months), probably high-overtone pulsators (added to the GCVS 9*

Observation Key	
★	Bright stars
▦	Mixed amplitudes
⌗	Mixed periods
◉	CCD or PEP

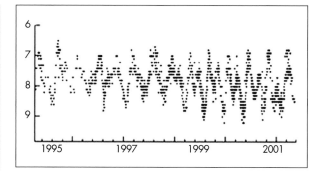

Figure 4.9. Light curve of the SRB-type variable star, Z UMa. Annual dates are indicated along the horizontal axis. Data provided by the VSNET. Used with permission.

July 2001 [The 76ᵗʰ Name List of Variable Stars - IBVS 5135]). GCVS

As you are probably beginning to notice, variability seems to be a fundamental characteristic of cool luminous stars. In fact, it is suggested that almost all late stars of spectral class M are variable at some level (Figure 4.9).

Semiregular variable stars possess some similarities to the Mira variables. The SRA and SRB stars are giants while the SRC are supergiants. The major difference between the SRA class and the Miras is that an SRA may have a visual light amplitude of less than $2^m.5$. In principle, the light curves can also be less regular than those of the Miras, however, Mira light curves can also be less regular. The SRB is similar to the SRA class but with a less obvious demonstration of period.

Like the Miras, SRA and SRB stars include spectral class types M, S and C.

SRC stars are generally thought to be massive with progenitors in excess of about 8 solar masses.

SRD stars are semiregular giants and supergiants of spectral type F, G or K. Occasionally, the spectra of theses stars will possess emission lines. SRD variables are generally considered poorly studied. A subgroup, discussed later in this chapter, is sometimes called UU Her stars.

SXPHE (SX Phoenicis stars)

– Phenomenologically, these variables resemble δ SCT variables and are pulsating subdwarfs of the spherical component, or old disk galactic population, with spectral types in the range A2–F5. They may show several simultaneous periods of oscillation, generally in the

range $0^d.04$–$0^d.08$, with variable amplitude light changes that may reach $0^m.7$ in V. Theses stars are present in globular cluster. **SXPHE(B)** *(subtype) not a recognized subtype within the GCVS. GCVS*

SX Phoenicis[14] variable stars are found within spectral types A2–F5 stars on the HR diagram. They display periods that are similar to the δ Scuti stars, however in comparison, their photometric amplitudes are greater, generally ranging from $0^m.3$ to $0^m.8$. Upon closer examination, metal abundance as well as space motion for these stars are typical of the Population II stars.

Population I and II classifications were originally made by Walter Baade to characterize two different groups of stars that are distinguished by their velocities (*kinematically distinct*), relative to the Sun. Population I stars posses velocities that are low when compared to Population II stars., relative to the Sun. Also, Population I stars are predominantly found in the disk of the galaxy whereas Population II stars are found above or below the disk.

Is has been suggested that SX Phoenicis stars are blue stragglers[15] in the post-main-sequence stage of evolution. The interesting thing about this suggestion is that one of the theories describing blue straggler evolution proposes that they are formed from coalescing binary stars; in other words, two stars that once orbited each other have come together to form a single star.

The discovery of SX Phoenicis stars among the blue stragglers in globular clusters marks the beginning of an exciting new age in the study of globular clusters. Like RR Lyrae stars and Cepheids, SX Phoenicis stars are useful as distance indicators, as probes of their respective stellar evolution phases, and for studying stellar population problems. Furthermore, since globular cluster SX Phe stars can be identified more easily than their field star counterparts, and the metal content and distances of the parent systems tend to be known, their relatively large numbers provide a basis for studying statistical relationships.

Observation Key

 Bright stars

 Mixed amplitudes

 Mixed periods

👁 CCD or PEP

---

[14]Named for "Phoenix" the mythical creature which, after being incinerated, grew again from its own ashes to fly away.

[15]Blue stragglers are a group of stars which show a reluctance to evolve off the main sequence due to some unusual aspect of their evolution. The cause may be mass exchange with a binary star or some process of internal chemical mixing that provides more hydrogen fuel in the core.

The light, color, and radial velocity curves of SX Phe stars are similar to those of RR Lyrae and Cepheid variables. At maximum light (phase = 0), the observed visual brightness is greatest, primarily because the star is hottest. The radius is increasing at the maximum expansion speed, having just been at maximum compression during the rise to maximum light. At minimum light, when the star is at its coolest, the radius of the star is at its minimum.

UUHER (*UU Herculis* stars)

– These stars are not recognized within the GCVS but have been the subject of study since 1928. UU Her stars are sometimes described as a subset of the semiregular D variables (SRD). The prototype star UU Her has been the subject of interest since its remarkable behavior was noticed in 1928. The suggestion has been made that this star switches between fundamental and first overtone modes.

As evidence, in 1899 and 1900 this star had a period close to 45 days, with an amplitude of about 1^m5. In 1901 the period changed to a much longer ~72 days, with an amplitude of 0^m8. This period persisted until 1905 when the star's period suddenly changed back to 45 days. This period remained stable until 1910 when the amplitude become so small that the star was labeled "invariable." Subsequent data from 1961 indicated a period of 45.6 days, and in 1984 a period of 71.6 days was indicated. Based upon the available information, it appears that UU Her switches periods between ~45 days and ~72 days.

In recent literature, three stars have been suggested as being classic members of this group: UU Her, 89 Her (V441 Her) and HD 161796. UU Her has often been listed among the RV Tauri stars but appears to now be listed as an SRD star. As you know, the hallmark of the RV Tauri stars is the alternating deep and shallow minima with a difference in depth being typically 0^m5–1^m. For UU Her, this phenomenon is present at a level of a few hundredths of a magnitude.

The star 89 Her shows bizarre behavior too. In 1977, a regular pulsation cycle with a period of ~64 days was evident. In 1978, the photometric pulsation suddenly changed to random fluttering that persisted through 1979. The regular pulsation cycle was reestablished in 1980.

When studied in 1983, HD 161796 seemed to possess two well-defined periods of 62 and 43 days, followed by an interval of non-variability. It was suggested that these two periods represent the fundamental and first overtone of radial pulsation. A study conducted in 1984 indicates that the amplitude of this star increased by about 50% and the star seemed to have brightened. Analysis of the data indicted a period of ~38 days with a second set of data indicating a period of ~54 days.

These stars are currently not listed as UU Her stars as this classification is provisional at best. Perhaps as more data is collected, a better understanding of these stars will develop.

ZZ (*ZZ Ceti stars*)

*– These are non-radially pulsating white dwarfs that change their brightness with periods from 30 s to 25 min and amplitudes from $0^m.001$ to $0^m.2$ in V. They usually show several close period values. Flares of 1 magnitude are sometimes observed; however, these may be explained by the presence of close UV Ceti companions. These variables are divided into the following subtypes: **ZZA** (subtype) – ZZ Cet-type variables of DA spectral type (ZZ Cet) having only hydrogen absorption lines in their spectra. **ZZB** (subtype) – ZZ Cet-type variables of DB spectral type having only helium absorption lines in their spectra. **ZZO** (subtype) – ZZ Cet type variables of the DO spectral type showing He II and C IV absorption lines in their spectra.* **GCVS**

Observation Key	
	Bright stars
	Mixed amplitudes
	Mixed periods
👁	CCD or PEP

ZZ Ceti stars are classed as pulsating stars, however, it is accepted that they cannot be radial pulsating stars because their periods are too long. In fact, multicolor observations have confirmed that the pulsation modes are non-radial g-modes. Usually, several periods are simultaneously excited and their frequencies are often split into close pairs by the slow rotation of the star.

The periods are usually considered to be extremely stable but unstable periods do occur, with period changes occurring over a few hours, and are probably caused by interactions of various periods. This phenomenon is generally described as the beating of closely spaced frequencies.

Single white dwarfs have masses near 0.6 solar masses

The *GCVS* lists approximately 30 ZZ Ceti-type variables with a handful occurring within novalike or dwarf nova systems. Most of these stars belong to the ZZA type.

Chapter 5
Cataclysmic (Explosive and Novalike) Variables

These are variable stars showing outbursts caused by thermonuclear burst processes in their surface layers (novae) or deep in their interiors (supernovae). We use the term "novalike" for variables that show novalike outbursts caused by rapid energy in the surrounding space volume (UG type stars) and also for objects not displaying outbursts but resembling explosive variables at minimum light by their spectral (or other) characteristics. The majority of explosive and novalike variables are close binary systems, their components having strong mutual influence on the evolution of each star. It is often observed that the hot dwarf component of the system is surrounded by an accretion disk formed by matter lost by the other, cooler, and more extended component.

GCVS

There are almost 800 *cataclysmic variables* (CVs) identified within the *General Catalog of Variable Stars*. The group with the most members within this class are the *dwarf novae* (U Geminorum type) that include the subtypes: UGSS, UGSU and UGZ. The dwarf novae (DN) are the longest-known and best-studied cataclysmic variables. They are observed extensively by amateur astronomers because of their recurring fast rise to brightness. To some extent, professionals rely upon amateurs to notify them when one of these stars goes into "outburst," the sudden brightening for which these variables are known. Professional astronomers are not able to constantly check the large number of dwarf novae for outbursts that can appear at largely, unpredictable times.

A *classical nova* (CN) represents a very different phenomenon from a dwarf nova. In this case the optical brightness increases by up to 20 magnitudes and by definition, only one such outburst is ever witnessed. The recurrence time of a classical nova is estimated to be in the range between 3300 and 100,000 years.

Another type of CV is the *recurrent nova* (RN) that displays outbursts on a time-scale of decades. RNs appear intermediate between the dwarf novae and classical novae. A type of star closely associated are the *symbiotic* (ZAND) stars, in which the accreting star is usually a main sequence star rather than a white dwarf. The various types of outbursts found within the group of recurrent novae are thought to be powered by different mechanisms depending upon the nature of the accreting star, that is, by accretion events onto main sequence stars or thermonuclear outburst onto white dwarf stars. More on accretion disks in a moment.

Novalike (NL) *stars* are named for their spectral and photometric similarities to the classical novae and dwarf novae; however, these CVs have had no observed outburst. Many novalike stars look like dwarf novae in permanent outburst, that is they are luminous disk-accreting systems. Other novalike stars are magnetic systems that have long-duration high and low states. The subtype known as VV Sculptoris stars have high and low states but do not have the classical signatures of a magnetic field.

Supernovae are different in the sense that they become brighter than normal only once and are generally of no interest to amateur astronomers after-wards. However, the discovery of a new supernova is cause for great excitement and joyous fanfare within the amateur community. If you are fortunate enough to detect one of these very rare and usually faint objects you'll understand the jubilation. You'll probably also become addicted to the search for these fantastic objects.

Novae do not possess the absolute brightness of supernovae but can display a greater apparent bright-ness making them easier to observe. Novae can appear in the sky where there was apparently no star visible or perhaps a well-known star suddenly becomes much brighter. They also cause great excitement within the variable-star community.

The other stars of this classification are not well observed by amateurs, are not well defined or only a small number of them are known to exist. Additional study is certainly indicated. Cataclysmic variable stars present some of the best challenges for amateur astronomers and the study of these interesting stars will, without a doubt, improve your observational skills as well as your understanding of complex stellar behavior.

In regards to complex stellar behavior, two important astrophysical phenomena are better understood within the context of cataclysmic variable stars: *Roche lobes* and *accretion disks*. Both phenomena are complex and of great interest to astrophysicist since they are basic processes found among many observed objects.

At least half of all the stars that we see in the sky are actually multiple systems, consisting of two or more stars in orbit about their common center of mass. In most of these systems the stars are sufficiently far apart that they have negligible influence upon one another. Essentially, both stars evolve independently, living out their lives in isolation except for the mild influence of gravity that gently binds them together within a common system.

If the two stars are very close, with a separation roughly equal to the diameter of the larger star, then one or both stars will have the outer layers of their atmosphere deformed by gravity, forming a teardrop shape. As a star rotates through the deformed bulge, called a *tidal bulge*, raised by its partner's gravitational pull, the star is forced to pulsate. These pulsations are stifled by various processes that eventually dissipate the orbital and rotational energies until a synchronous rotation and a circular orbit are achieved. Once this process is complete, the same side of each star always faces the other as the system rotates rigidly in space and no further energy can be lost by tidally driven pulsation. In a situation like this, the distorted star may even lose some of its photospheric gases to its companion (Figure 5.1).

As one of the stars evolves within a binary system, it will expand to fill successively larger volumes of space. You recall how evolving stars "push" their atmospheres farther away from the core. During this long evolutionary process, the demeanor of a binary star system is influenced by the volume of space it fills. For example, widely separated binary stars are nearly spherical and this type of situation describes a *detached binary* in which both stars evolve nearly independently.

A different situation exists if one star expands sufficiently. Then its atmospheric gases can escape

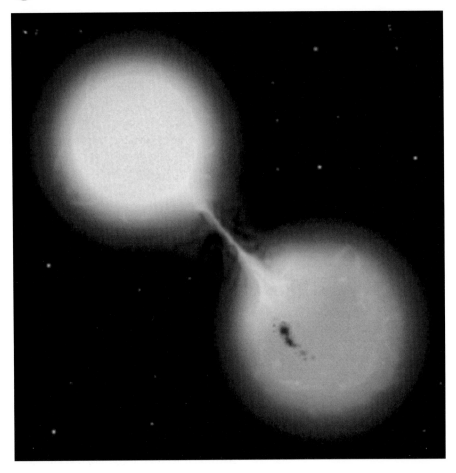

from the gravitationally influenced volume of space surrounding the star and will be drawn toward its companion by gravitational forces. The location at which these escaping gases move from the teardrop-shaped volume of space surrounding one star to the other, is called the *inner Lagrangian point*. The teardrop-shaped volumes of space surrounding binary stars, in which stellar gas remains close to the star, are called *Roche lobes*.[1] The transfer of mass from one star to the other can begin when one of the stars has expanded beyond its Roche lobe and the mass begins flowing through the inner Lagrangian point. Such a system is called a *semidetached binary*. The star that fills its Roche lobe and loses mass is usually called the

Figure 5.1. Artist's conception of a binary system demonstrating the concept of Roche lobes. Copyright: Gerry A. Good.

---

[1] The term "Roche lobe" is named in honor of the nineteenth-century French mathematician Edouard Roche.

secondary star and its companion, the star receiving the material, is called the primary star. The primary star may be either more massive or less massive than the secondary star.

In some situations both stars fill, or even expand beyond, their Roche lobes. Then the two stars share a common atmosphere bounded by a dumbbell-shaped volume of space. Such a system is called a *contact binary* because of the contact between their shared atmospheres.

The orbital motion of a semidetached binary can prevent the gases that escape from the swollen secondary star from falling directly onto the primary star. The primary's orbital motion is usually sufficient to constantly move it out of the path of the gases that spill through the inner Lagrangian point. If the radius of the primary star is less than about 5% of the binary separation, the gas stream will miss striking the primary's surface. When this happens, the gas stream assumes an orbit around the primary and forms an accretion disk of hot gas in the orbital plane.

Roche lobes and accretion disks are of great importance when studying cataclysmic variables because it's impossible to study them in a laboratory. Many of these stars produce both as a result of their geometry.

AM Canis Venaticorum stars are helium-rich, ultra-short-period CVs. These systems are presumably composed of two white dwarfs.

AM Herculis stars possess strong magnetic fields and are known as "polars." These stars, actually binary systems, possess extremely strong magnetic fields that dramatically affect the accretion disk surrounding the white dwarf of the binary system.

DQ Herculis stars are known as "intermediate polars" because, as with AM Herculis stars, they have strong magnetic fields that dominate the accretion flow.

ER Ursae Majoris stars, also known as *RZ LMi* stars, are dwarf nova that exhibit both a standstill phase similar to *Z Cam* stars and short outbursts that resemble normal outbursts of *SU UMa* stars.

Novae are interacting binary systems consisting of a white dwarf and a cool dwarf star that produce sudden outbursts gradually returning to normal over several months, years or decades. Within this book, the novae classification has been divided into two groups for easier study. Novalike stars are considered separately from novae because of their ambiguous nomenclature.

Supernovae are rare stellar explosions that result in extreme outbursts, increasing the star's brightness by 20^m or more. In the end, the star is destroyed or dramatically changed.

TOADS are also known as Tremendous Outburst Dwarf Novae and their true existence is in debate. Certainly they exist as a celestial object but their unambiguous nature is in question and the argument suggests that perhaps they do not deserve to be distinguished from certain other CVs.

U Geminorum stars are also known as *dwarf novae* (DN) and are generally believed to be semidetached binary systems containing a white dwarf and a low-mass main sequence star. The Roche-overflow gas from the secondary star forms an accretion disk around the compact object. The chief source of the visual light of DNs is from the accretion disks.

The group of stars known as *V Sagittae* stars are classified as novalike, cataclysmic variables, but are not recognized within the *GCVS*. They do not generally fit into any of the patterns established for novalike CVs.

W Sagittae stars are distinguished from most DNs by their unusually large-amplitude outbursts. It is not clear that the *WZ Sge* stars can be unambiguously classified as different from *SU UMa* stars.

Z Camelopardalis stars have intermediate mass-transfer rates and are believed to display phases showing dwarf nova-type activity when the disk is thermally unstable and standstills when the disk is thermally stable, similar to novalike variables. Table 5.1 lists the *GCVS* classifications.

AMHER (AM Herculis stars)

– Usually considered a subdivision of the novalike cataclysmic variables, AM Her stars are also known as "polars." These binary systems contain a synchronously rotating, magnetized white dwarf and a cool companion that is near the main sequence. The accretion occurs towards the magnetic poles and is the reason for the name polar. The stars show polarized optical radiation, strong X-ray radiation, short period modulation and long-term bright and low states with orbital periods less than 3.5 hours. **not recognized within the GCVS**

Observation Key	
★	Faint stars
	Small amplitudes
	Short periods
👁	CCD or PEP

The exotic star AM Herculis is the namesake of the group of cataclysmic variables known as "polars," a

Table 5.1. Cataclysmic variable stars arranged in alphabetical order by designation

Variable type		Designation (and subclasses)	
AM Canis Venaticorum		**AMCVN**	helium-rich, ultra-short-period CVs
AM Herculis	*	**AMHER**	CVs with strong magnetic field, known as "polars"
DQ Herculis	*	**DQHER**	magnetic, fast spinning CVs known as "intermediate polars"
Novae		**N** (four subclasses indicated below)	
		NA	fast novae
		NB	slow novae
		NC	very slow novae
		NR	recurrent novae (three subclasses indicated below)
		T Pyx stars	
		U Sco stars	
		T CrB stars	
Novalike		**NL**	
		RW Tri	
		UX UMa	
SW Sextantis		**SW Sex**	
TOADS		**TOADS**	
VY Sculptoris	*	**VYSCL**	anti-dwarf nova stars
Supernovae		**SN** (two subclasses indicated below)	
		SNI	type I supernovae
		SNII	type II supernovae
U Geminorum		**UG** (three subclasses indicated below)	
(dwarf novae)		**UGSS**	SS Cyg stars
		UGSU	SU UMa stars
		UGZ	Z Cam stars
WZ Sagittae		**WZSGE**	
V Sagittae	*	**VSGE**	V Sagittae stars
Z Andromedae		**ZAND**	**S**ymbiotic systems

* Indicates a designation found within the literature but not recognized within the *GCVS*.

class of cataclysmic variables in which the magnetic field of the primary star (white dwarf) completely dominates the accretion flow of the system. AM Her was discovered in 1923 by M. Wolf in Heidelberg, Germany during a routine search for variable stars. The star was listed in the *GCVS* as an irregular variable with a range of brightness from $12^{m}.0$ to $14^{m}.0$. AM Her remained listed as an irregular until 1976, when the true nature of the star was finally understood.

The discovery of AM Herculis introduced a new class of highly magnetic stars to the group of cataclysmic variables. AM Her stars display magnetic field strengths so powerful that it prevents the formation of an accretion disk around the white dwarf (Figure 5.2) and locks the two stars of the binary system together so

they always present the same face to each other. The white dwarf star spins at the same rate as the two stars orbit each other, a *synchronous rotation* that is the defining characteristic of an AM Her star.

The light curve of AM Her (Figure 5.3) appears to have the temperament of a super-violent tornado itself. There is apparently more than one source of radiation wreaking havoc on the star. The variations in AM Her may be thought to belong to two groups, the long-term changes and the short-term changes; characterized by the existence of two different states, one the *active* or "on" state, in which the luminosity fluctuates around $13^m.0$, and the other *inactive* or "off" state, where the brightness remains at about $15^m.0$. These two states are thought to be the result of active and inactive mass-transfer rates from the secondary to the primary star.

Figure 5.2. Artist's conception of a polar CV showing the white dwarf's magnetic field capturing the accreting material. Copyright: Gerry A. Good.

DQHER (DQ Herculis stars)

– Another group of stars usually considered a subdivision of the novalike cataclysmic variables, DQ Her stars are also known as "intermediate polars." These binary systems contains a non-synchronously rotating, magnetized white dwarf and a cool companion that is near the main sequence. Accretion occurs in the outer regions and forms a disk. Close to the white dwarf, the accretion disk is disrupted by the magnetic field and mass flows by way of an accretion column towards the star's magnetic

Observation Key	
★	Faint stars
〽	Small amplitudes
☾	Short periods
◉	CCD or PEP

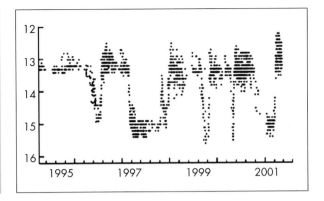

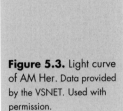

Figure 5.3. Light curve of AM Her. Data provided by the VSNET. Used with permission.

poles. *Light variations are caused by eclipse effects and by the rotationally modulated accretion effects. Several intermediate polars have undergone nova explosions.* **not recognized within the GCVS**

Although not officially listed as a group of variable stars within the *GCVS*, astronomers recognize DQ Her stars as an important group of variable stars. DQ Her stars are also known as intermediate polars (IP) and display strong magnetic fields surrounding the white dwarf star of the binary system.

An accretion disk has formed around these stars, but it is disrupted close to the primary (white dwarf) star because of the magnetic field. In the case of intermediate polars, the magnetosphere is not strong enough to synchronize the orbits of the rotating white dwarf with the orbital period of the system, such as is observed in AM Her stars.

The *Intermediate Polar Homepage*, a Web site dedicated to the study of these stars, can be found at <*http://lheawww.gsfc.nasa.gov/users/mukai/iphome/iphome. html*>. You will find a list of intermediate polar candidates and basic information pertaining to the stars, including where to obtain finder charts and references.

ER UMa (ER Ursae Majoris stars, or RZ Leo Minoris stars)

– ER Ursae Majoris stars are dwarf nova subtypes that exhibit a standstill like phase similar to Z Camelopardalis stars. They exhibit short outbursts that resemble

normal outbursts of SU Ursae Majoris stars in regards to their rapid decline rate. **not recognized within the GCVS**

ER UMa was originally discovered as an ultraviolet-excess object and confirmed to be a cataclysmic variable in 1986. Until very recently the object had been little studied. It was only in 1992 that the dwarf nova nature of this object was first noticed. Intensive visual observations have revealed that the object varies between $12^{m}\!.3$ and $15^{m}\!.2$ in V. The most peculiar photometric behavior of this object is the presence of a *standstill-like phase* following the initial decline of some outbursts.

The star stays in this phase for ten days or so at $0^{m}\!.5$ to $1^{m}\!.0$ fainter than the outburst maximum. These characteristics closely resemble short standstills observed in Z Cam stars.

Another intriguing feature is the presence of short, but occasionally bright, outbursts with a maximum decline rate of $0^{m}\!.7$ per day. These short outbursts resemble normal outbursts of SU UMa stars in the sense of its rapid decline rate, although no previously known SU UMa stars have been shown to exhibit the Z Cam-type characteristics at the same time.

Observation Key	
★	Faint stars
⌁	Small amplitudes
☻	Short periods
◉	CCD or PEP

N (Novae)

– Close binary systems with orbital periods from $0^{d}\!.05$ to $230^{d}\!.$ One of the components of these systems is a hot dwarf star that suddenly, during a time interval from one to several dozen or several hundred days, increases it brightness by $7^{m}\!-19^{m}$ in V, then returns gradually to its former brightness over several months, years, or decades. Small changes at minimum light may be present. Cool components may be giants, subgiants, or dwarfs of K–M type. The spectra of novae near maximum light resemble A–F absorption spectra of luminous stars at first. Then broad emission lines (bands) of hydrogen, helium, and other elements with absorption components indicating the presence of a rapidly expanding envelope appear in the spectrum. As the light decreases, the composite spectrum begins to show forbidden lines characteristic of the spectra of gaseous nebulae excited by hot stars. At minimum light, the spectra of novae are generally continuous or resemble the spectra of Wolf–Rayet stars. Only spectra of the most massive systems show traces of cool components.

Observation Key	
★	Mixed stars
⌁	Large amplitudes
☻	Mixed periods
◉	Visual, CCD/PEP

*Some novae reveal pulsations of hot components with periods of approximately 100^s and amplitudes of about $0^m.05$ in V after an outburst. Some "novae" eventually turn out to be eclipsing systems. According to the features of their light variations, novae are subdivided into fast (NA), slow (NB), very slow (NC), and recurrent (NR) categories. **NA** (subgroup) – Fast novae displaying rapid light increases and then, having achieved maximum light, fading by 3^m in 100 or fewer days. **NB** (subgroup) – Slow novae that fade after maximum light by 3^m in 150^d. Here the presence of the well-known "dip" in the light curve of novae similar to T Aur and DQ Her is not taken into account: The rate of fading is estimated on the basis of a smooth curve, its parts before and after the "dip" being a direct continuation of one another. **NC** (subgroup) – Novae with a very slow development and remaining at maximum light for more than a decade, then fading very slowly. Before an outburst, these objects may show long-period light changes with amplitudes of 1^m–2^m in V; cool components of these systems are probably giants or supergiants, sometimes semiregular variables, and even Mira variables. Outburst amplitudes may reach 10^m. High excitation emission spectra resemble those of planetary nebulae, Wolf–Rayet stars, and symbiotic variables. The possibility that these objects are planetary nebulae in the process of formation is not excluded. **NR** (subgroup) – Recurrent novae, which differ from typical novae by the fact that two or more outbursts (instead of a single one) separated by 10–80 years have been observed. GCVS*

Together with dwarf novae and novalike variables, *novae* are interacting binary stars generally possessing a short orbital period. Novae consist of a massive white dwarf, the primary, and a cool dwarf star, the secondary. The secondary star overflows its Roche lobe and thereby loses mass to the primary star. This matter forms an accretion disk around the primary star and is finally accreted onto its surface. Instabilities in the accretion disk lead to short and long period photometric variability at the stage of minimum light.

The cause of the nova outburst is a thermonuclear runaway reaction that occurs in the accreted hydrogen-rich layer near the surface of the massive white dwarf into which C and O nuclei from the outer layers of the white dwarf are mixed. When the critical pressure is reached, hydrogen burning via the CNO cycle begins within the degenerate hydrogen-rich outer layer. A

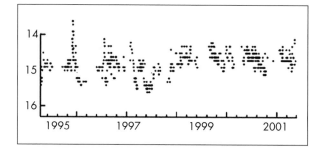

Figure 5.4. Light curve of Q Cyg (NA). Data provided by the VSNET. Used with permission.

rapid increase of the temperature leads to a lifting of the degeneracy and to the formation of a shock wave. This, in combination with radiation-driven mass-loss, produces an expanding atmosphere of large size and of high absolute magnitude, typically $M_v = -6$ to -9, at maximum light. Decreasing mass loss with an ongoing energy release causes a decline of the visual light output, a shrinking of the photosphere, and radiative heating of the ejected material, resulting in interesting spectroscopic phenomena in the course of the outburst (Figure 5.4).

The detailed light and spectral properties of novae are complex, and depend on the white dwarf mass and its chemical composition, and mixing of CO-rich nuclei into the accreted material as well as on dust formation in the ejected shell. Each nova has its own unique, characteristic photometric and spectroscopic evolution. In spite of this, novae can be broadly classified into several subgroups.

NA – fast novae which, after maximum light, decline three magnitudes in visual light in 100 days or less. They usually have fairly smooth light curves and generally have higher absolute magnitudes (see Figure 5.5).

NB – slow novae which decline three magnitudes from maximum in visual light in more than 100 days. They have usually fairly structured light curves and, as a rule, fainter absolute magnitudes.

NC – very slow novae, which remain near maximum light for years or even decades. The bulk of these objects are symbiotic stars, accreting objects with late-type giant companions. They are often called symbiotic novae.

NR – recurrent novae that show repeated outbursts at time intervals of decades. Recurrent novae are usually fast novae, often have giant companions,

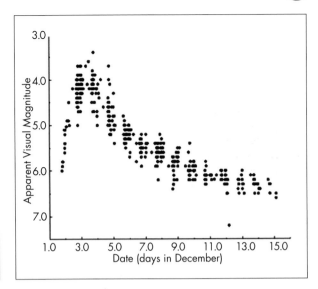

Figure 5.5. Light curve of V1494 Agl (Nova). Chart provided by the AAVSO. Used with permission.

and the accreting white dwarfs are probably near the Chandrasekhar limit, a position that allows explosions in degenerate material under high pressure.

Members of the groups NA and NB are also called *classical novae*. Their absolute magnitude is correlated with the speed of the light-curve decay: faster novae are more luminous at maximum light. The fastness is measured by t_2- or t_3-time: a nova takes a certain time,

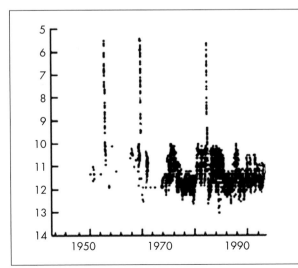

Figure 5.6. Light curve of RS Oph. Data provided by the VSNET. Used with permission.

measured in days, to decline by 2 or 3 magnitudes from maximum light.

When first detected, novae are always a surprise and produce much excitement within the professional and amateur community. Recent nova discoveries include: Nova Sgr 2001 No. 3 = V4740 Sgr that reached a maximum brightness of ~6^{m}8, Nova Sgr 2001 No. 2 = V4739 Sgr that reached a maximum brightness of ~6^{m}4 and Nova Cyg 2001 No. 2 = V2275 Cyg that reached a maximum brightness of ~6^{m}6. These three novae were all within the observing capabilities of binocular observers or small telescopes.

NL (Novalike variables)

– Variable stars which are insufficiently studied objects resembling novae by the characteristics of their light changes or by spectral features. This type includes, in addition to variables showing novalike outbursts, objects with no burst ever observed; the spectra of novalike variables resemble those of old novae, and small light changes resemble those typical for old novae at minimum light. However, quite often a detailed investigation makes it possible to reclassify some representatives of this highly inhomogeneous group of objects to other types. **GCVS**

Observation Key	
	Mixed stars
	no amplitude
	no period
	Visual, CCD/PEP

The novalike group of variable stars is classified as *insufficiently studied* stars. These stars include stars with no observed bursts. They are included in this group because the spectra resemble old novae at minimum light. Quite often a detailed investigation makes it possible to reclassify some representatives of this highly inhomogeneous group into some other type of variable star.

We do not know exactly how novae look for the long interval between outbursts. According to the *hibernation hypothesis*, accretion can be dramatically reduced and novae may then not have the appearance of a novalike star. If the accretion rate and the magnetic field strength of the white dwarf are low enough then quasi-periodic disk instabilities can occur and the object will be classified as a *dwarf nova*. If the white dwarf is massive enough, greater than 0.6 solar mass, nova explosions can occur, and the object is classified as a *nova* if such an event has occurred in the last few decades and was properly recorded. In all other cases, that is when signatures of accretion on the white dwarf via a disk or an accretion column are present in the spectrum and the object cannot be clearly classified as N or DN, it is classified as NL.

SN (Supernovae)

Observation Key

★ Mixed stars
▨ Large amplitudes
🔺 Single event
👁 Visual, CCD/PEP

– Stars that increase, as a result of an outburst, their brightness by 20^m and more, then fade slowly. The spectrum during an outburst is characterized by the presence of very broad emission bands, their widths being several times greater than those of the bright bands observed in the spectra of novae. The expansion velocities of SN envelopes are in the thousands of km/s. The structure of a star after outburst alters completely. An expanding emission nebula results and a (not always observable) pulsar remains at the position of the original star. According to the light curve shape and the spectral features, supernovae are subdivided into types I and II. **SNI (subtype)** *– Type I supernovae. Absorption lines of Ca II, Si, etc., but not hydrogen lines are present in the spectra. The expanding envelope almost lacks hydrogen. During 20–30 days following maximum light, the brightness decreases by approximately $0^m.1$ per day, then the rate of fading slows and reaches a constant value of $0^m.014$ per day.* **SNII (subtype)** *– Type II supernovae. Lines of hydrogen and other elements are apparent in their spectra. The expanding envelope consists mainly of H and He. Light curves show greater diversity than those of type I supernovae. Usually after 40–100 days since maximum light, the rate of fading is $0^m.1$ per day.* **GCVS**

The stellar explosions called supernovae are one of the rarest and most spectacular phenomena observed in the Universe. Of course, *gamma-ray bursts* (GRBs) produce more energy but they are not yet widely observed by amateur astronomers and the actual mechanism responsible for their enormous release of energy is still debated. When you consider the total energy that is released during a supernova and the terminal results, supernovae are the most dramatic stellar events observed frequently by amateurs.

In the year 1885, a star appeared near the nucleus of M31, the Great Andromeda galaxy, and reached an apparent magnitude of 7.2, about 1/10th as bright as the nebula itself. The outburst aroused considerable interest but, since the nature and distance of the Andromeda nebula were unknown, the discussions were largely speculative, and were eventually forgotten.

In 1917, two much fainter novae were found on photographs of M31 taken at Mount Wilson. Interest in the subject was immediately revived. The nebula was repeatedly photographed, and during the next five or six years, a total of 22 faint novae was assembled by various observers at Mount Wilson. These novae formed a compact, homogeneous group with fairly uniform maxima. The star of 1885 stood out as a striking exception with a maximum several thousand times brighter than those of the faint group.

The faint novae proved to be comparable with the common or normal novae while the star of 1885 together with the similar outbursts of fainter nebulae were provisionally assigned to a new group of extremely bright novae. The star of 1885, for instance, had evidently reached a maximum of 100 million suns. The differences between normal novae and bright novae seemed so pronounced that, quite early, they were provisionally assumed to represent distinct types. By 1933, a nomenclature had been generally adopted, and henceforth the bright novae were called super-novae.

A supernova is a rare type of stellar explosion which dramatically changes the structure of a star in an irreversible way. Large amounts of matter are expelled at high velocities. The light curve in the declining part is powered by thermalized quanta, released by the radio-active decay of elements produced in the stellar collapse, mainly ^{56}Co and ^{56}Ni. The ejected shell interacts with the interstellar medium and forms a supernova remnant (SNR), which can be observed long after the explosion in the radio, optical and X-ray regions.

Early detailed study of supernovae led to the recognition, by Rudolph Leo Bernhard Minkowski,[2] of two groups (I and II) which differed radically in their spectra and, according to Wilhelm Heinrich Walter

---

[2]Minkowski joined the Mount Wilson Observatory staff in 1936. He studied spectra, distributions, the motions of planetary nebulae and headed the National Geographic Society–Palomar Sky Survey that photographed the entire northern sky in the 1950s.

Baade,[3] in their maximum luminosities and the patterns of their light curves as well.

SN I have fairly similar light curves and display a small spread in absolute magnitudes. Spectra around maximum show absorption lines of Ca II, Si II and He I, but lack lines of hydrogen. They occur in intermediate and old stellar populations. Their progenitor stars are not clearly identified, but massive white dwarfs that accrete matter from a close companion and are pushed over the Chandrasekhar limit are good candidates.

Supernovae are vast explosions in which a whole star is blown up. Most are seen within distant galaxies as "new" stars appearing close to the galaxy of which they are members. They are extremely bright, rivaling, for a few days, the combined light output of all the rest of the stars in the galaxy.

As most observed supernovae occur in very distant galaxies they are too faint even for the largest telescopes to be able to study them in great detail. Occasionally they occur in nearby galaxies and then a detailed study in many different wavebands is possible. The last supernova to be seen in our galaxy, the Milky Way system, was seen in 1604 by the famous astronomer Kepler. The brightest since then was supernova 1987A in the Large Magellanic Cloud, a small satellite galaxy to the Milky Way. The brightest supernova in the northern sky for 20 years is supernova 1993J in the galaxy M81 which was first seen on March 26 1993.

Supernovae fall into two types whose evolutionary history is different. Type I supernovae result from mass transfer inside a binary system consisting of a white dwarf star and an evolving giant star. Type II supernovae are, in general, single massive stars which come to the end of their lives in a very spectacular fashion.

Type I supernovae are even brighter objects than those of type II. Although the explosion mechanism is somewhat similar the cause is rather different. The origin of a Type I supernova is an old, evolved binary system in which at least one component is a white dwarf star. White dwarf stars are very small compact stars which have collapsed to a size about one tenth that of the Sun. They represent the final evolutionary stage of all low-mass stars. The electrons in a white dwarf are subject to quantum mechanical constraints (the matter is called degenerate) and this state can only be maintained for star masses less than about 1.4 times that of the Sun.

The pair of stars loses angular momentum until they are so close together that the matter in the companion star is transferred into a thick disk around the white dwarf and is gradually accreted by the white dwarf. The mass transferred from the giant star increases the mass of the white dwarf to a value significantly higher than the critical value whereupon the whole star collapses and the nuclear burning of carbon and oxygen to nickel yields sufficient energy to blow the star to bits. The subsequent energy released is, as in the Type II case, from the radioactive decay of the nickel through cobalt to iron.

The structure of all stars is determined by the battle between gravity and radiation pressure arising from internal energy generation. In the early stages of a star's evolution the energy generation in its center comes from the conversion of hydrogen into helium. For stars with masses of about 10 times that of the Sun this continues for about ten million years. After this time all the hydrogen in the center of such a star is exhausted and hydrogen "burning" can only continue in a shell around the helium core. The core contracts under gravity until its temperature is high enough for helium burning, into carbon and oxygen, to occur. The helium burning phase also lasts about a million years but eventually the helium at the star's center is exhausted and it continues, like the hydrogen burning, in a shell. The core again contracts until it is hot enough for the conversion of carbon into neon, sodium and magnesium. This lasts for about 10 thousand years.

This pattern of core exhaustion, contraction and shell burning is repeated as neon is converted into oxygen and magnesium (lasting about 12 years), oxygen goes to silicon and sulfur (about 4 years) and finally silicon goes to iron, taking about a week. No further energy can be obtained by fusion once the core has reached iron and so there is now no radiation pressure to balance the force of gravity. The crunch comes when the mass of iron reaches 1.4 solar masses. Gravitational compression heats the core to a point where it endothermically decays into neutrons. The core collapses from half the Earth's diameter to about 100 kilometers in a few tenths of a second and in about one second becomes a 10 kilometer diameter neutron star. This releases an enormous amount of potential energy primarily in the form of neutrinos that carry 99% of the energy.

A shock wave is produced which passes, in about 2 hours, through the outer layers of the star causing fusion reactions to occur. In concert, the two processes form the heavy elements. In particular the silicon and sulfur, formed shortly before the collapse, combine to give radioactive nickel and cobalt which are responsible for the shape of the light curve after the first two weeks.

When the shock reaches the star's surface the temperature reaches 200 thousand degrees and the star explodes at about 15,000 kilometers/sec. This rapidly expanding envelope is seen as the initial rapid rise in brightness. It is rather like a huge fireball which rapidly expands and thins allowing radiation from deeper in towards the center of the original star to be seen. Subsequently most of the light comes from energy released by the radioactive decay of cobalt and nickel produced in the explosion.

TOADS (Tremendous outburst amplitude dwarf novae)

Observation Key

 Mixed stars
 Large amplitudes
 Mixed periods
CCD or PEP

*– In 1995, Howell, Szkody and Cannizzo distinguished a particular type of dwarf novae, characterized by the very large outburst amplitudes of their optical outbursts (6 to 10 magnitudes) and very long intervals between the outbursts (months to decades). These TOADs are a subset of the SU UMa systems (dwarf novae that show both "normal outbursts" and "super-outburst"). Apart from the very long intervals and the very large amplitudes, TOADs also differ from the other SU UMa systems in that almost all TOAD outbursts are super-outbursts. **not recognized within the GCVS**

With such an intriguing name, TOADs *must* be interesting objects. Like all CVs, they are members of semidetached binary systems that have orbital periods ranging from approximately one to twelve hours. When studying dwarf novae, distinctions are made using orbital periods. Those stars with orbital periods below the CV period gap (~2–3 hours), belong to the SU UMa group.

Astronomers Steve Howell and Paula Szkody have launched a multiyear project to study the fainter CVs with periods below this gap. Their research has discovered many binary systems that share some of the characteristics of the SU UMa systems, but that have additional unique properties.

These dwarf novae have been named TOADs because of their unusually large outburst amplitudes. Whether or not these systems are really the same thing as SU UMa stars at different stages of their evolution or they are a separate subclass of dwarf novae is still to be determined. TOADs and SU UMa systems can both be classified as dwarf novae with orbital periods less than ~2.5 hours. Both types of systems exhibit super-outbursts.

The unique traits of TOADs are: longer, brighter and less frequent super-outbursts, likely low mass transfer rate (less than 10^{-11} solar masses/year), possibly very low viscosity disk material (10–100× below normal) at minimum, accretion disks may be advected, secondary stars may be degenerate.

There is much debate regarding these stars with suggestions made that they are simply a subset of the SU UMa stars. Dr. Joe Patterson of Columbia University, runs the *Center for Backyard Astrophysics*, a group of amateur astronomers dedicated to the observation of various cataclysmic variable stars. I have found his comments interesting:

There has been much mention of TOADs and I thought it was worth entering a dissent on this term, which seems to me is astronomically and zoologically poor. Dwarf nova fans are familiar with the term SU UMa stars and the meaning is clear: dwarf novae whose eruptions divide very distinctly into long and short, and which show super-humps in their long eruptions. Some people also use the term WZ Sge stars to refer to the subset of SU UMa stars that either: (a) show the longest outburst intervals, (b) have few or no short outbursts, or (c) qualify according to both (a) and (b). Because this is somewhat vague and because Nature provides no dividing line, many of us either do not use the term WZ Sge stars or use it only as a convenient shorthand for a more cumbersome phrase, the most infrequently erupting SU UMa stars. Whether it is a useful subclass is harder to say.

UG (U Geminorum variable stars)

*– Quite often called dwarf novae. They are close binary systems consisting of a dwarf or subgiant K–M star that fills the volume of its inner Roche lobe and a white dwarf surrounded by an accretion disk. Orbital periods are in the range of $0^{d}\!.05$–$0^{d}\!.5$. Usually only small, in some cases rapid, light fluctuations are observed, but from time to time the brightness of a system increases rapidly by several magnitudes and, after an interval of from several days to a month or more, returns to the original state. Intervals between two consecutive outbursts for a given star may vary greatly, but every star is characterized by a certain mean value of these intervals, i.e. a mean cycle that corresponds to the mean light amplitude. The longer the cycle, the greater the amplitude. These systems are frequently sources of X-ray emission. The spectrum of a system at minimum is continuous, with broad H and He emission lines. At maximum these lines almost disappear or become shallow absorption lines. Some of these systems are eclipsing, possibly indicating that the primary minimum is caused by the eclipse of a hot spot that originates in the accretion disk from the infall of a gaseous stream from the K–M star. According to the characteristics of the light changes, U Gem variables may be subdivided into three types: **UGSS (subgroup)** – SS Cygni type variables. They increase in brightness by 2^{m}–6^{m} in V in 1^{d}–2^{d} and in several subsequent days return to their original brightness. The values of the cycle are in the range 10 days to several hundred. **UGSU (subgroup)** – SU Ursae Majoris type variables. These are character-ized by the presence of two types of outbursts called "normal" and "super-maxima." Normal, short outbursts are similar to those of UGSS stars, while super-maxima are brighter by 2^{m}, are more than 5 times longer (wider), and occur three times less frequently. During super-maxima the light curves show superposed periodic oscillations (super-humps), their periods being close to the orbital ones and amplitudes being about $0^{m}\!.2$–$0^{m}\!.3$ in V. Orbital periods are shorter than $0^{d}\!.1$; companions are dM spectral type. **UGZ (subgroup)** – Z Camelopardalis type stars. These also show cyclic outbursts, differing from UGSS variables by the fact that sometimes after an outburst they do not return to the original brightness, but during several cycles retain a magnitude between*

maximum and minimum. The values of cycles are from 10 to 40 days, while light amplitudes are from 2^m to 5^m in V. GCVS

Dwarf novae outbursts are intrinsically much less luminous events than classical novae outbursts. Their peak absolute magnitudes are at least 100 times weaker. Dwarf novae are known to recur, with some recurring on time-scales as short as a few weeks. Dwarf novae also have short durations, lasting a few days. Dwarf novae can also exhibit a variety of unusual behaviors. SU UMa type sources occasionally exhibit extremely long outbursts known as superoutbursts. Z Cam stars will occasionally get stuck in standstills during which their brightness is both below outburst stage and well above quiescent levels. VY Scl stars, also known as anti-dwarf novae, will spend most of their time in an outburst state, with occasional dips into quiescence that last for a few days. Finally, there are novalikes which behave much like novae long after their eruptions, but which have never exhibited novae outbursts. They are also distinct from dwarf novae outbursts in that they have permanently high rates of mass transfer.

The principal source of electromagnetic radiation in a dwarf nova system is the accretion disk. The companion star to the white dwarf is a low-mass red dwarf star filling its Roche lobe with matter streaming onto the accretion disk through the inner Lagrange point. The gas stream from the L1 point impacts the

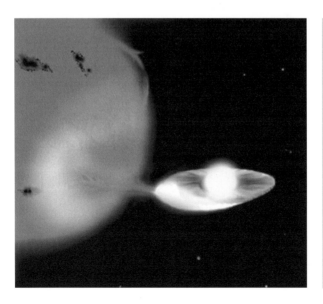

Figure 5.7. Artist's conception of a dwarf novae demonstrating the formation of an accretion disk. Copyright: Gerry A. Good.

accretion disk and creates a hot spot. Matter gradually transports through the accretion disk onto the surface of the white dwarf, generating temperatures which make the disk much hotter and brighter than either star. The dwarf nova outburst and other related phenomenon are believed to be caused by variations in the accretion rates through the disk. Material reaching the white dwarf surface through the disk must pass through a violent transition region, called the boundary layer: it is here that the X-rays in dwarf novae originate. This is shown dramatically by the recent observations of X-ray eclipses in HT Cas; the eclipse duration is the same as that of the white dwarf as determined by optical observations. The sharpness of the transitions into and out of the eclipse proves that the X-ray emitting region has a size comparable to that of the white dwarf.

Dwarf novae are generally believed to be semidetached binaries containing a white dwarf and low-mass main sequence stars. The Roche-overflow gas from the secondary star forms an accretion disk around the compact object. The chief source of the visual light of CVs is from the accretion disks. Several sorts of disk instabilities dramatically affect the luminosity of the disk, and then become observable as a variation in the optical flux. This feature not only provides us with one of the best opportunities in direct investigation of the physics of accretion disks through observations of CVs and XTs, but also enables us to reveal the nature of specific objects of astrophysical importance by applying the known physics of accretion disks.

Dwarf novae show semiperiodic outbursts with a typical amplitude ranging from 2 to 6 magnitudes, and with a recurrence period of 10 to 1000 days. In contrast, novalike variables do not show prominent outburst activities.

Dwarf novae are further subclassified according to their light behavior: SS Cyg stars display approximately regular, recurring outbursts, Z Cam stars display "standstills" during which the stars show little variation at brightness between maxima and minima, SU UMa stars display two distinct types of outbursts, short (normal) outbursts and super-outbursts that are brighter than the normal outbursts.

As stated in the first section, the visual light of CVs for the most part reflects the energy output from the accretion disk, hence the cause of variation should primarily be sought in the accretion process itself.

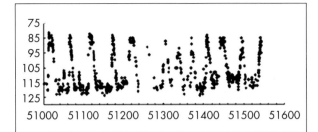

Figure 5.8. Light curve of U Gem (dwarf nova). Data provided by the VSNET. Used with permission.

Among several mechanisms to explain this rich variety of light variation of CVs, only two of them seems to have remained viable: mass-transfer instability and disk instability. The former paradigm primarily assumes that the changing mass-transfer rate from the secondary produces the luminosity variation of the accretion disk; the latter, in contrast, does not assume change of mass-transfer rate, but the intrinsic instability of the accretion disk produces temporal changes in mass-accretion rate in the disk which is observed as quiescent and outburst states. The discrimination of these two paradigms in variable accreting system has been, whether explicitly or implicitly posed, always one of the main goals of both observers and theoreticians.

After a long period of debate, a fairly good consensus in ordinary CVs seems to have been reached between most observers and theoreticians concerning the natural explanation of dwarf nova phenomenon: the disk instability model. The basic disk instability idea explains the dwarf nova phenomenon in the following way: the disk accumulates the accreted mass during quiescence and accretes it to the white dwarf during outburst. The nature of the disk instability which triggers such interchange of disk status was not known at that time. Subsequent theoretical studies finally discovered the thermal instability of the accretion disk due to the partial ionization of the hydrogen. This thermal instability has been shown to not only successfully reproduce the various light curves of SS Cyg-type dwarf novae but also gives a natural explanation of two basic types of CVs: dwarf novae and novalike variables.

The difference between dwarf novae and novalike variables is explained in the scheme of thermal instability theory, in that the higher mass-transfer rate in novalike variables produce thermally stable accretion disks.

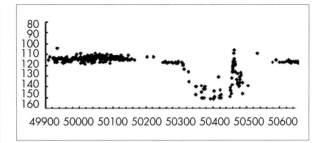

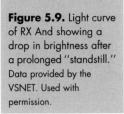

Figure 5.9. Light curve of RX And showing a drop in brightness after a prolonged "standstill." Data provided by the VSNET. Used with permission.

The Z Cam stars have intermediate mass-transfer rates, and are believed to share properties of these two subclasses, that is, phases showing dwarf nova-type activity when the disk is thermally unstable, and standstills when the disk in thermally stable like novalike variables.

VSGE (V Sagittae[4] stars)

Observation Key

 ★ Mixed stars
Mixed amplitudes
Mixed periods
Visual, CCD/PEP

– Stars of this type are often classified as novalike cataclysmic variables but, in fact, they do not fit into any of the patterns established for this class. The nature of these stars is still not clear, but there seems to be a consensus in the literature that they are binary systems with an evolved component. The nature of this component, however, has not been clearly established. Ideas involving a subdwarf, white dwarf, neutron star,

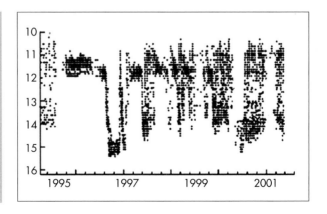

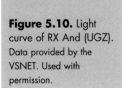

Figure 5.10. Light curve of RX And (UGZ). Data provided by the VSNET. Used with permission.

---

[4]Sagitta (pronounced SAH-jit-a), the Arrow, is recognized as the arrow Hercules shot eastwards, still in flight, and also as Cupid's arrow.

and a black hole plus a He (helium) main-sequence star have been considered. **not recognized within the GCVS**

The variable star V Sagittae has defied classification despite much attention from photometric observers since its discovery in 1902. It has often been classified as a novalike cataclysmic variable but, in fact, does not fit into any of the patterns established for this class.

V Sge was discovered in 1902 as a variable star but it wasn't until 1965 that it was discovered to be a binary system with an orbital period of 12^h3. Since then it has remained a puzzle among the variable stars. There are a small number of stars, however, that show properties quite similar to those of V Sge.

The nature of these stars is still not clear but there seems to be a consensus in the literature that they are binary systems with an evolved component. The nature of this component, however, has not been clearly established. Ideas involving a subdwarf, white dwarf, neutron star, and a black hole along with an He main sequence star have been considered for V Sge.

V Sge is the first object of its class to be identified and studied in detail. It has been shown that this star is a double spectroscopic binary. V Sge is also a faint and soft Einstein[5] source. The long-term photometric behavior of this system is based on observations made over a period of 70 years. It has been shown that an optical brightness variation exists, including high and low states separated by up to 2^m0. A semiregular period of $\sim240^d$ has also been claimed.

It is important to establish the basic structure and evolutionary status of this new class of binaries; in particular, the nature of the compact star is of fundamental importance to the understanding of these objects. To this point there has been no agreement in the literature of the nature of the compact, more evolved, component of the binary system.

The hypothesis of a white dwarf as the compact star has been discussed as one of the possibilities for V Sge, the other two possibilities being a neutron star or a black hole. The most popular model for explaining the supersoft X-ray sources seems to be hydrostatic nuclear burning on the surface of a white dwarf. This can occur when a massive white dwarf accretes at rates approach-

---

[5]The second of NASA's three High Energy Astrophysical Observatories, HEAO-2, renamed Einstein after launch.

ing 10^{-6} solar masses per year. The progenitor stars should be in the range of 6–8 solar masses. According to this idea, about 100–200 such objects are expected to exist in the Galaxy.

The neutron star hypothesis has also been raised for V Sge stars. It can be shown that under some circumstances, accretion onto a neutron star may produce a supersoft X-ray binary.

VYSCL (VY Sculptoris[6] variable stars)

<table>
<tr><td colspan="2" align="center">**Observation Key**</td></tr>
<tr><td></td><td>Mixed stars</td></tr>
<tr><td></td><td>Small amplitudes</td></tr>
<tr><td></td><td>Mixed periods</td></tr>
<tr><td>👁</td><td>Visual, CCD/PEP</td></tr>
</table>

– The VY Scl stars, sometimes called anti-dwarf novae, are cataclysmic variables, the light curves of which can be characterized by occasional drops from steady high states into low states lasting up to several hundred days. These low states probably result from episodes of low mass transfer from the companion star. **not recognized within the GCVS**

The study of the Balmer radial velocity curve of the cataclysmic variable VY Scl has revealed a new value for the orbital period (0^d232), which significantly differs from the previously accepted one. It also shows a gamma modulation that is compatible with the presence of a third component, indicating that this object could be a hierarchical triple system. The CV is a low-inclination ($i \sim 30°$) system with a massive white dwarf. There are two possibilities for the nature of the hypothetical third component: either a = 0.8 solar mass star or a low-luminosity higher mass star (probably a compact object). The value for the period of the third object's orbit would be $\sim 5^d8$, which makes the triple system dynamically stable.

WZSGE (WZ Sagittae stars)

– The WZ Sge subclass of dwarf novae are distinguished from most dwarf novae by their unusually large-amplitude outbursts (6^m–9^m) which last much longer

---

[6]Sculptor, the sculptor was originally named by Lacaille as the Sculptor's Workshop or Studio.

but recur less frequently than the "normal" outbursts of dwarf novae. It is not, however, clear that the WZ Sge stars are distinct from the SU UMa[7] subclass of dwarf novae[i]. **not recognized within the GCVS**

WZ Sagittae, the prototype star for this class of variable star, is a cataclysmic variable which had outbursts in 1913, 1946, 1978, and 2001. WZ Sge is an eclipsing binary with a period of ~ 82^m.0. In 1969, the star was originally cataloged as a recurrent nova. In 1976, its apparently low luminosity led to a suggestion that it is more closely related to dwarf novae than to other recurrent novae. Observations of the 1978 outburst confirmed its resemblance to dwarf novae.

The item of interest here is the fact that the outburst light curve of WZ Sge is different from those normally observed in dwarf novae.

Observation Key

★ Mixed stars
▦ Large amplitudes
☺ Mixed periods
◉ Visual, CCD/PEP

ZAND (Z Andromedae[8] variable stars)

– Symbiotic variables of the Z Andromedae type. They are close binaries consisting of a hot star, a star of late type, and an extended envelope excited by the hot star's radiation. The combined brightness displays irregular variations with amplitudes up to 4^m in V. A very inhomogeneous group of objects. **GCVS**

Observation Key

★ Mixed stars
▦ Large amplitudes
☺ Mixed periods
◉ Visual, CCD/PEP

Z And stars, also known as *symbiotic stars*, are interacting binary systems. The defining characteristic of this diverse group is that, in addition to erratic photometric variability, their spectra simultaneously show spectral signatures such as the molecular absorption features of a cool giant star. Studies over limited wavelength regions have often resulted in symbiotic stars being incorrectly classified as something else, most often as peculiar planetary nebulae. The molecular absorption features are frequently only present in infrared spectra.

The giant component of the binary system is usually of spectral type M or C. There are also a few so-called

---

[7] A subclass of the U Geminorum type variable (dwarf novae) characterized by two types of outbursts called "normal" and "supermaxima" and by superposed "superhumps" during a super-maximum outburst.

[8] In Greek mythology, Andromeda was the daughter of King Cepheus and Queen Cassiopeia.

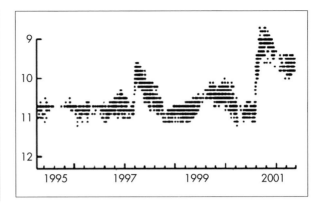

Figure 5.11. Light curve of Z And (ZAND). Data provided by the VSNET. Used with permission.

yellow symbiotics which have G-type spectra. The other star in the binary system may be a low-mass main sequence star or compact object, such as a subdwarf star, white dwarf or neutron star. The interaction that results in the *symbiotic phenomenon*, including erratic variability and high-excitation emission lines, begins when mass is transferred from the giant star to its partner. Most of the well-studied systems contain either a main sequence star which accretes by direct tidal overflow from the giant, or a white dwarf which accretes from the giant's stellar wind. Many Z And stars show evidence of an accretion disk. The transfer of mass will often produce a hot spot within the accretion disk. In many cases this hot spot provides the temperature necessary for ionizing part of the circumstellar environment and producing the emission lines. Symbiotic stars are closely related to the even rarer VV Cep systems within which a late-type supergiant interacts with an O or a B star.

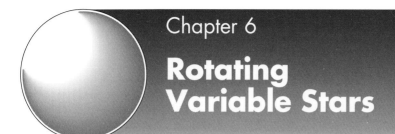

Chapter 6

Rotating Variable Stars

Variable stars with nonuniform surface brightness and/ or ellipsoidal shapes, whose variability is caused by axial rotation with respect to an observer. The nonuniformity of surface brightness distributions may be caused by the presence of spots or by some thermal or chemical inhomogeneity of the atmosphere caused by a magnetic field whose axis is not coincident with the rotation axis.

GCVS

There are over 900 stars classified as rotating variables within the *General Catalog of Variable Stars*. The α^2 Canum Venaticorum stars are the most numerous, followed by the BY Draconis variables. Only a dozen or so of the FK Coma Berenices type variables are known. As a group, these variables do not receive a great amount of attention from amateur astronomers, probably because they possess small amplitudes and are almost impossible to accurately observe without the aid of instruments. Generally, the range of brightness between maximum and minimum for rotating variable stars is on the order of hundredths to tenths of a magnitude. On the other hand, when observed with the proper instruments these stars will generally reward the patient observer with rapid amplitude changes (periods) on the order of hours in some cases.

Obviously, every star rotates but if the star also has relatively large permanent or semipermanent surface features, similar to sunspots, then the star will appear to vary in brightness, or even color, as the rotation of the star carries the surface features across the observer's line of sight. These change may be so subtle as to not be visually detectable but can be detectable using CCDs or PEP methods. It doesn't take much of a difference in temperature since luminosity varies with temperature raised to the fourth power. This means that a small temperature change, as in the case of a cool "spot" on a star, results in a huge change in luminosity. For this reason, sunspots look black, even though their temperature is many thousands of degrees. The star will not vary of course, if the axis of the rotation points to the observer, or if the surface features are symmetric about the axis of rotation.

As a class, rotating variable stars will provide you with an excellent opportunity to improve your observational abilities using instruments and in the process, provide a valuable service to science. Observing rotating stars, collecting good data, and plotting their light curves requires attention to detail and the application of rigorous observing methods.

α^2 Canum Venaticorum stars are main sequence stars (luminosity class V) of spectral types B8p-A7p and they display strong magnetic fields. The "p" in their nomenclature indicates a *peculiar* chemical composition because their spectra show abnormally strong lines of Si, Sr, Cr, and rare earths that vary with rotation, magnetic field, and brightness changes. These light variations occur over a range from half a day to more than 160 days. The amplitudes of the light changes are usually in the range of $0^m01–0^m1$ in V. The best-known group of rotating variables are the peculiar A stars, known as Ap stars. In these stars, the surface features (spots) are caused by a strong magnetic field locking cool chemical structures into relatively stable positions within the stellar gases. The variations in brightness and color are usually small, but in one or two cases they are almost large enough to be seen visually.

BY Draconis stars are usually described as a subset of the classical UV Ceti flare stars. BY Dra stars are commonly described as displaying low amplitudes with periods of a few days, dK or dM spectral type and emission lines of Ca II. You will find within the literature that some astronomers believe that all flare stars are subject to BY Dra-type variability from time to time. To verify this hypothesis will require very careful photometric observations conducted over a period of years. Some BY Draconis stars display amplitudes near 0^m5, easily within the observational capabilities of dedicated amateur astronomers using instruments.

Table 6.1. Rotating variable stars arranged in alphabetical order by designation

Variable type	Designation	
α^2 Canum Venaticorum	**ACV**	B8p–A7p stars showing small variations due to large "starspots" generated by intense magnetic fields.
	ACVO	Rapidly oscillating α^2 CVn variables.
BY Draconis	**BYDRA**	dKe–dMe stars showing quasi-periodic light changes
Ellipsoidal variable	**ELL**	rotating ellipsoidal variables stars. May be any spectral class
FK Comae	**FKCOM**	G–K III stars that are rapidly rotating, spotted, and in some cases they are binary systems
Pulsars	**PSR**	optically variable pulsars
SX Arietis	**SXARI**	B0p–B9p stars with intense He I and Si III lines

Rotating *ellipsoidal variable* stars are close binary systems that display changes in brightness by periods equal to their orbital motion. Because the two stars are close together, gravity deforms these stars' atmospheres sufficiently to cause variability. Light amplitudes do not exceed $0^m\!.1$ in V but this is easily detectable using CCD or PEP methods.

FK Comae stars are giant stars (luminosity class III) of spectral type G–K, with rapid rotational velocities. They also display irregular surface features that cause fluctuating brightness. One theory regarding the formation of these stars is that they are coalesced binary systems, perhaps evolved W UMa stars (see Chapter 7, "Close Binary Eclipsing Systems").

Optically variable pulsars are rapidly rotating neutron stars with strong magnetic fields, radiating in the radio, optical, and X-ray regions. Pulsars emit narrow beams of radiation and periods of the light changes coincide with rotational periods (from $0^s\!.004$ to 4^s). The amplitudes of the light pulses may reach $0^m\!.8$.

SX Arietis stars are sometimes called helium variables. Periods of light and magnetic field changes coincide with rotational periods, while amplitudes are $\approx 0^m\!.1$ in V. These stars are usually described as high-temperature analogs of the ACV variables. The *GCVS* classifications are listed in Table 6.1.

ACV (α^2 Canum Venaticorum stars)

– These are main-sequence stars with spectral types B8p–A7p and displaying strong magnetic fields. Spectra

show abnormally strong lines of Si, Sr, Cr, and rare earths whose intensities vary with rotation, magnetic field, and brightness changes ($05-160^d$ or more). The amplitudes of the light changes are usually in the range of $0^m.01-0^m.1$ in V. **ACVO** *(subtype) – Rapidly oscillating α^2 Canum Venaticorum variables. These are nonradially pulsating, rotating magnetic variables of Ap spectral type. Pulsation periods are in the range 6^m-12^m (0004–001), while amplitudes of light variation caused by the pulsation are about $0^m.01$ in V. The pulsational variations are superposed on those caused by rotation.* **GCVS**

α^2 Canum Venaticorum stars, also known as Ap^1 and $roAp^2$ are stars in which the surface is severely depleted of helium (He) with, at the same time, overabundance of iron (Fe), silicon (Si) and chromium (Cr) in spots. These stars have been known since the early days of spectral classification, when the phenomenon was first detected.

Of interest when attempting to unambiguously classify these stars, the discrimination made in the *GCVS* in regards to ACV and SXARI stars seems irrelevant; one classification is often assigned to a star for which the other one would be more appropriate.

These chemically peculiar (CP) stars, in general, are stars that possess spectral signatures of chemical peculiarities such as strongly enhanced spectral lines of iron and rare-earth elements. In this group there is a *magnetic sequence*, referring to stars that demonstrate a strong, global magnetic field. This does not mean that the HgMn stars, or metallic-line (Am) stars, etc. have no magnetic field at all; however, stars designated within the nonmagnetic sequence may exist without a magnetic field or exist with a significantly weaker global effect or even with a strong field of complicated structure, such that the measurable effect, averaged-out over the visible disk, is insignificant. Ap stars have global surface magnetic fields in the order of 0.3 to 30 kg,[3] and their effective magnetic-field strength varies with rotation, a situation that led to an interpretation in terms of the *oblique-rotator model* in which the

Observation Key	
★	Faint stars
	Small amplitudes
	Short periods
👁	CCD or PEP

---

[1]The "p" suffix indicates that the star is chemically peculiar, in which the spectra reveal chemical signatures such as Fe and rare-earth elements.

[2]The "ro" prefix indicates rapidly oscillating Ap stars.

[3]Thousands of times stronger than that of the Sun.

magnetic axis is oblique to the rotation axis.[4] The time-scales of light variations seen in Ap stars range from minutes to decades.

Ap stars are intrinsically slow rotators but the hotter stars rotate faster than the cooler ones. The length of the rotation period can be derived by plotting their spotted surface variations that change as the star rotates. Most periods are of the order of one day to one week, with a tail towards longer periods. Other sources of variability, such as binary motion or pulsation, may be superposed, so careful analysis is required.

A source of further information regarding chemically peculiar stars in general, and specifically α^2 Canum Venaticorum variables, is the *Peculiar Newsletter*, of the IAU Working Group on Ap and Related Stars. This is a specialized astronomical circular founded in Vienna, Institut d'Astronomie de l'Université de Lausanne, CH-1290 Chavannes-des-Bois, Switzerland *<http://obswww.unige.ch/~north/APN/Welcome.html>*.

Of particular interest may be Issue No 24, October 17, 1995, which contains the *Table of Ap and Am Stars* in order of variable-star names. The table includes not only the Ap and Am stars quoted as variable in the 4th edition of the *General Catalog of Variable Stars* but also those for which a variable-star name has more recently been assigned (name-lists Nos. 67–72). Also, each star of the *Catalog général des étoiles Ap et Am* having received a variable-star name is listed.

BY (BY Draconis[5] stars)

Observation Key

 Mixed stars

 Small amplitudes

Long periods

CCD or PEP

– These stars are emission-line dwarfs of dKe–dMe spectral type showing quasiperiodic light changes with periods from a fraction of a day to 120^d and amplitudes from several hundredths to $0^m\!.5$ in V. The light variability is caused by axial rotation of a star with a variable degree of nonuniformity of the surface brightness (spots) and chromospheric activity. Some of these stars also show flares similar to those of UV Ceti stars, and in these cases they also belong to the latter type and are simultaneously considered eruptive variables. GCVS

---

[4]This model was first suggested by Stibbs in 1950.
[5]Draco, the Dragon, identified as the beast that guarded the golden apples of the garden of the Hesperides, and was killed by Hercules when he came to fetch the apples as his eleventh labor.

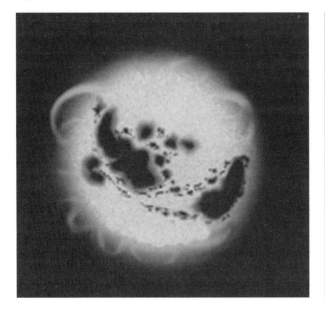

Figure 6.1. Artist's conception of a BY Draconis- type variable showing the large starspots responsible for this type of star's variability. Copyright: Gerry A. Good.

The BY Draconis variables are dKe and dMe stars. Their spectral designation shows that they are late (K–M) dwarfs (prefix "d") that exhibit hydrogen line emissions (suffix "e") in their spectra. The variability in these stars is produced by axial rotation with irregular surface brightness (Figure 6.1). A region of cool spots localized on one hemisphere of the star causes the irregular light variations.

BY Dra variables are one of the many stars showing interesting activity within their chromosphere. A cursory examination of these stars may lead you to believe that they are similar to RS CVn binaries, however, BY Dra variables can be either binary or single. This physical characteristic was used to prove that a binary configuration is not directly responsible for this type of chromospheric activity.

Several of the BY Dra variables also show UV Ceti-type flares and the *GCVS* recognizes this phenomenon and has added the UV classification to some of these stars.

A few stars classified BY Dra within the *GCVS* probably should be classified as FK Com stars instead. Though undoubtedly single, spotted, and varying as a result of rotational modulation, they are not the proper spectral type and/or luminosity class for BY Dra variables. Examples would be OP And (gK1), V390 Aur (K0 III), EK Eri (G8 IV-III), and V491 Per (G8 IV).

The variable star BY Dra, the group's prototype, was discovered in 1966. At that time, star spots were used to explain its variable nature. Another BY Dra star, YY Gem, was discovered in 1926, but was classified as an eclipsing binary. It is nevertheless clearly a BY Dra variable; its spectral type is dMe+dM2e and its variability between the eclipses was correctly identified as a starspot wave. The *GCVS*, however, emphasizing its eclipses and its flares, classifies it EA+UV.

ELL (Rotating ellipsoidal variable stars)

Observation Key

⭐ Mixed stars
📈 Small amplitudes
🌗 Mixed periods
👁 CCD or PEP

– These are close binary systems with ellipsoidal components, which change combined brightnesses with periods equal to those of orbital motion because of changes in emitting areas toward an observer, but showing no eclipses. Light amplitudes do not exceed $0^m.1$ in V. GCVS

Ellipsoidal variable stars are by definition binary systems. Only in response to gravity are the shapes of these stars distorted sufficiently to cause variability. You may pause to consider the binary nature of these systems and ponder the chances of an eclipse occurring. Certainly, when you have two or more stars there is some non-zero probability that one may by chance pass in front of another, relative to the observer.

However, to be considered an ellipsoidal variable, this must not occur. *Eclipses are disallowed!* Although ellipsoidal variables are physically similar to eclipsing variables with out-of-eclipse light variations, ellipsoidal variables have small orbital inclinations that prohibit an eclipses (Figure 6.2).

A major difficulty in studying ellipsoidal variables has been the lack of a comprehensive catalog of these

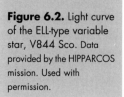

Figure 6.2. Light curve of the ELL-type variable star, V844 Sco. Data provided by the HIPPARCOS mission. Used with permission.

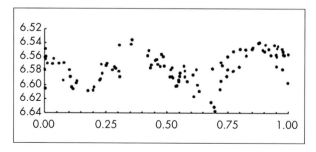

stars, making it difficult to study them as a class or to compare any particular variable with another ellipsoidal. Upon close examination of this group of variable stars, it is clear that the ellipsoidal classification of stars has been used as a convenient catch-all for variables with fragmentary or peculiar light curves. How fortunate for amateur astronomers! This is an excellent group of stars on which to conduct serious study.

Long-term, precise observations will be required to identify misclassified stars. Little reference material will be found regarding these stars. Charts are lacking. Comparison stars have not been identified. Analysis techniques are not well established. For the serious amateur, these stars invite your attention.

FKCOM (FK Comae Bernices[6] stars)

– These are rapidly rotating giants with nonuniform surface brightnesses, which have spectral types G–K with broad H and K Ca II emission and sometimes Ha. They may also be spectral binary systems. Periods of light variation (up to several days) are equal to rotational periods, and amplitudes are several tenths of a magnitude. It is not excluded that these objects are the product of further evolution of EW (W UMa) close binary systems. GCVS

Observation Key	
★	Mixed stars
	Small amplitudes
	Mixed periods
◉	CCD or PEP

FK Comae variables are rapidly rotating giant stars varying as a result of irregular surface brightness. A region of cool spots localized on one hemisphere of the star causes the irregular brightness. As originally defined, the class included late-type giants with very high rotation speed (short rotation period), evidence of extreme chromospheric activity, but displaying no evidence of large velocity variations (Figure 6.3). The GCVS allows binaries to be included in the class.

FK Comae itself rotates so rapidly that the most reasonable evolutionary scenario involves the coalescence of a W UMa-type binary and a surrounding optically thick spun-up envelope. Other stars assigned to this class do not rotate so rapidly and may be simply evolved single A-type stars, which have not lost much of

---

[6]Coma Berenices, the Hair of Berenice. Berenice is the wife of Ptolemy Euergetes, king of Egypt who vowed to sacrifice her hair if her husband was successful in waging war on the Assyrians.

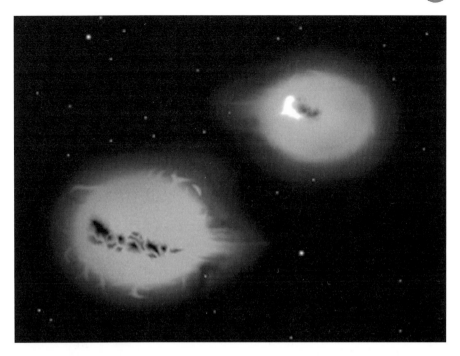

Figure 6.3. Artist's conception of a FK Com-type variable showing the fast orbiting nature and irregular surface brightness of these binary systems. Copyright: Gerry A. Good.

their original rapid main sequence rotation. If binaries are allowed in the class, then their rapid rotation will be a result simply of synchronization with a rather short orbital period.

Again, because we're observing light variation caused by spots rotating across the face of distant stars, the light amplitude for these stars is in the hundredths to tenths of a magnitude. Definitely work for instruments.

PSR (Optically variable pulsars)

Observation Key

★ Faint stars
Small amplitudes
Mixed periods
CCD or PEP

*– These stars are rapidly rotating neutron stars with strong magnetic fields, radiating in the radio, optical, and X-ray regions. Pulsars emit narrow beams of radiation and periods of the light changes coincide with rotational periods (from 0^s004 to 4^s), while amplitudes of the light pulses reach 0^m8. **GCVS***

Only a handful of pulsars are listed within the *GCVS* and all are extremely faint. As optical targets, these exotic objects cannot be recommended for casual examination or study. If considered for serious study, earnest amateurs will require large aperture telescopes augmented with sensitive instruments such as CCDs.

When you consider that amateur astronomers occasionally image gamma-ray bursts only hours old, report supernovae shining at a "bright" 18^m0, and exceed the "depth" of the 200 inch telescope on Mount Palomar using a 16 inch telescope and CCD, I hesitate to absolutely suggest that the study of pulsars is out of the question. With this caveat duly presented, these objects can certainly be examined through the efforts of others. After all, as in real life, we can't hunt whales or travel to the Moon, but we can read about it.

SXARI (SX Arietis[7] variable stars)

– These stars are main-sequence B0p–B9p stars with variable intensity He I and Si III lines and magnetic fields. They are sometimes called helium variables. Periods of light and magnetic field changes (about 1^d) coincide with rotational periods, while amplitudes are $\approx 0^m.1$ in V. These stars are high-temperature analogs of the ACV variables. GCVS

Observation Key

★ Mixed stars

▦ Small amplitudes

☺ Short periods

◉ CCD or PEP

Many years ago, these stars were called spectrum variables of type-A, silicon variables, helium variables, or Ap silicon stars. Now, the SX Arietis stars are usually described as high-temperature analogs of the α^2 Canum Venaticorum stars. As mentioned within the section on ACV stars, the discrimination made within the *GCVS* regarding the notations and SX Ari seems inconsistent.

Within the *General Catalog of Variable Stars*, you will find 33 *SX Ari* stars. Ten of these stars are classified as uncertain. Thirteen more are found within *The 67–73 Name List of Variable Stars (NL)*, and six in *The 74th NL*.

SX Ari variables show variations in brightness that can only be reliably detected with instruments. They are not good candidates for visual observation but provide those with instruments an excellent opportunity to observe a complete cycle in one evening. When observed with a CCD or by photoelectric means, their short periods allow a detailed study during a single observing season (Figure 6.4).

Observing techniques for these stars are the same as with the α^2 Canum Venaticorum stars. Long-term

---

[7]Aries, the Ram, associated with a Ram and given the title "prince of constellations" or "leader of the constellations."

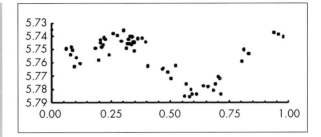

Figure 6.4. Light curve of the SXARI-type variable star, SX Ari. Data provided by the HIPPARCOS mission. Used with permission.

observation, using CCD or PEP techniques, is required and because of their short period, a complete cycle, perhaps more, should be the goal for an evening of work.

A Few Comments Regarding Low-Amplitude Stars

I want to take a few moments at the end of this chapter to discuss the importance of observing two classes of variable stars that we've examined: eruptive and rotating variable stars. The majority of these stars display small amplitudes, therefore they require instruments to properly study and as a result may be ignored by many amateur astronomers.

The need for instruments shouldn't be viewed as an obstacle or a reason to believe that the proper study of these stars is beyond the capabilities of amateur astronomers. On the other hand, the study of low-amplitude variable stars is certainly not mandatory. There is no rule that requires visual observers to purchase the equipment needed to study these stars. As you should understand by now, there are a sufficient number of variable stars to keep the visual observer busy for many lifetimes providing valuable information. Visual observers should feel no pressure to begin using instruments.

With that said, what I would like to point out is that many amateur astronomers with instruments, meaning CCDs and photometers, are observing well-studied stars to a level not really necessary and could better use their precious time and turn their equipment toward the study of these poorly observed stars. I've noticed that within several variable star databases the brightness estimates for many fast, low-amplitude variable stars are spaced days apart. I want to be clear on this point so let me say it another way; a large number of these low-amplitude stars display fast changes in amplitude, on the order of minutes to hours, but the estimates reported are taken once every few days and so completely miss the important characteristics of the star. This kind of reporting fails to provide the kind of information needed by other astronomers and will not provide you with the necessary information base to conduct meaningful analysis of your own data. When you make reports like this you are failing to optimize your time. Fast variables, those experiencing changes in brightness over the course or minutes to hours, need to be monitored frequently, usually every few minutes throughout a complete cycle if possible. Think of it this way, if you wanted to understand the annual temperature changes for your geographical location, you wouldn't record the air temperature once a month and then predict the temperature extremes for the whole year, would you? Or check the stock index once a month to determine which to buy or sell? Of course you wouldn't. Don't do this when observing variable stars.

Chapter 7

Close Binary Eclipsing Systems

We adopt a triple system of classifying eclipsing binary systems: according to the shape of the combined light curve, as well as to physical and evolutionary characteristics of their components. The classification based on light curves is simple, traditional, and suits the observers; the second and third classification methods take into account positions of the binary-system components in the (Mv, B–V) diagram and the degree of inner Roche lobe filling. Estimates are made by applying the simple criteria proposed by Svechnikov and Istomin (1979).

GCVS

Within the *GCVS* you will find over 6000 eclipsing binary variable stars. The *Algol-type* eclipsing binary stars are the most numerous; however, over one thousand eclipsing stars have not been well studied and as a result have not been categorized within the three major groups (i.e. EA, EB, or EW). One thousand ill-defined eclipsing binaries, classified as "E:", can certainly be described as a target-rich environment for any observer!

Of course, upon close examination most of these stars are considered to be faint, even at maximum brightness, or they display a small amplitude and as a result will require instrumentation to properly observe them throughout their entire cycle. If you're a serious visual observer, this does not present a hopeless challenge. A casual examination of the variables classified as "E:" readily expose a handful of stars that you can easily study by using visual methods. For example, consider the following two stars: HO CMa, maxima – 7^m.55, minima – 8^m.62 with a period that seems to be undetermined; and V536 Mon, maxima – 9^m10, minima – 10^m10 with a period of 31^d035. Both of these stars can be studied using visual means and without a doubt many others with similar characteristics can be found.

If you question that visual observers are able to adequately observe stars with similar, apparently limiting characteristics such as a small amplitude or faintness, read carefully within this chapter the results obtained by Kari Tikkanen, an observer from Finland who uses binoculars to observe variable stars. His results are encouraging.

While eruptive, pulsating, cataclysmic and rotating stars are said to be intrinsic variables because their variability is caused by different internal physical mechanisms, eclipsing binaries are called extrinsic variables and their study requires complex physical models in addition to the requisite stellar physics to describe their variable characteristics. Constructing these physical models requires the use of both astrophysics and geometry; good reasons that the study of eclipsing binary stars is considered a complex labor. Let's examine some of the physical properties that have some bearing on the study of these stars.

Usually, astronomers classify physical binary stars according to the manner of detection. To understand what we can learn from binary stars, it helps to understand the different methods used to observe them.

Visual binaries are physical pairs in which both members can be resolved with your eye, a telescope or a camera. Over 65,000 visual binary stars have been studied by astronomers. Should you ever become tired of variable stars, visual binary stars provide a great area of study for amateur astronomers!

With *spectroscopic binaries*, the individual stars cannot be resolved. The orbital motions are revealed by periodic Doppler shifts in the star's spectral lines. There are two subtypes of spectroscopic binary stars: those in which one spectrum can be detected; and those in which two sets of spectral lines are seen. The latter, displaying lines of both stars, yield more information to the observer.

An *eclipsing binary* is a pair of stars with a mutual orbit that is seen nearly edgewise and therefore, is producing eclipses. Because our line of sight lies in, or nearly in, the orbital plane, the stars alternately pass in front of each other. The light curves that you prepare after observing these stars will reveal much about the pair. Eclipsing binary stars is the topic of this chapter.

An *eclipsing-spectroscopic binary* shows both Doppler shifts and detectable eclipses. This is the most informative type of binary, permitting very detailed analysis of motions, masses, and sizes of stars.

An *astrometric binary* is one revealed not by Doppler shifts or eclipses, but by motions measured with respect to background stars.

Conducting eclipsing binary studies often involves the combination of photometric and spectroscopic data. Photometric data is principally light curves while spectroscopic data is primarily radial velocity curves produced by measuring the Doppler shift in spectral lines. In principle, interrogation of the light curve surrenders the orbital inclination and eccentricity, relative stellar sizes and shapes, perhaps the mass ratio in a few cases, the ratio of surface brightnesses, and brightness distributions of the stars among other quantities. If radial velocities are available, the masses and semimajor axis may also be determined. Radial velocity can be retrieved from journal articles so you can use it to carry out very thorough studies of these stars. Many other parameters describing the system and the individual stars may be determined, in principle, if the light curve data displays high precision and the stars do not differ greatly from your assumed model. A great computer program that will help you study eclipsing binary stars is *Binary Maker 2.0*, provided by David H. Bradstreet, Contact Software, Norristown, PA 19401–5505.

In some cases it's possible to determine the position of each star within the binary system, especially during the eclipse. The greatest loss of brightness is when the fainter star passes in front of the brighter star, causing the total brightness of the system to drop. When the fainter star is positioned off to one side of the brighter star, relative to the direct line of sight, the system is brightest. As the fainter star passes behind the brighter star, the system again loses light but not as much as when a portion of the brighter star's light is blocked. As we will see, there are many configurations involving two stars (Figure 7.1).

Shown next are prototypical light curves corresponding to the classical categories of Algol, β Lyrae, and W Uma stars, known as EA, EB and EW variables respectively.

Algol type (EA) light curves are typically almost flat-topped, suggesting that any photometric effects due to

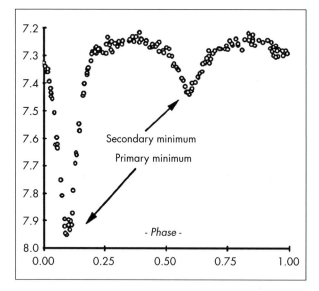

Figure 7.1. Light curve illustrating the minimum and maximum brightness for an eclipsing binary.

the proximity of the stars are small. A large difference between the depths of the two minima is evident and in some cases the smaller of the two is difficult to detect. You may even find, at some wavelengths (for example, when using science filters), that the secondary minimum may be undetectable and there may even be an increase in light near the expected phase of secondary minimum due to the *reflection effect*. Within a binary system, the presence of a second star leads to an increased brightness on the side that faces toward the companion star. The increased brightness is caused by heating from the radiant energy of the companion star. Obviously, this results in an increase of temperature. As you probably understand, since thermal energy is the physical cause for this phenomenon it is somewhat misleading to use the expression *reflection effect* (Figure 7.2).

One effect of reflection on binary star light curves is to increase the light around the secondary eclipse relative to that near the primary eclipse. Another effect is to produce a concave, or upward curvature, of the light curve between eclipses. When the two stars within a binary system have similar temperatures and are close but not actually over-contact, it may be necessary to consider multiple refection effects. The eclipsing binary BF Aurigae is an example for such a binary. The first star heats the second star, and the second star, now warmer, then heats the first star more than otherwise expected because of its own raised temperature. This

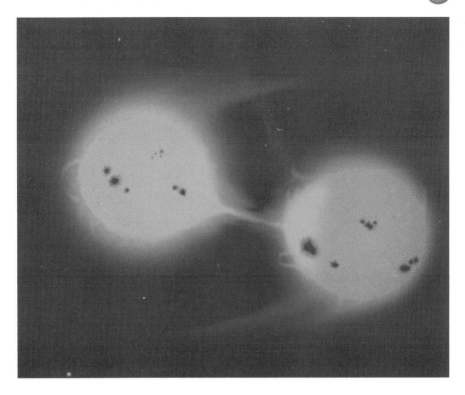

Figure 7.2. Artist's conception of the reflection effect. Notice the brightness of the facing sides of each star. Copyright: Gerry A. Good.

process is iterative, meaning that it compounds itself, and leads to higher temperatures on the facing hemispheres of the two stars.

Beta Lyrae (EB) light curves, on the other hand, show continuous variability, characteristic of tidally distorted stars with a large difference in depths of minima. This type of light curve usually indicates stars with different surface brightnesses. The prototype variable (β Lyrae) for this group of eclipsing stars was discovered to be a variable star by John Goodricke in 1784. The changes are most easily noticed by comparing β to its neighbor γ Lyrae, which has a magnitude of 3^m2. At maximum light, β and γ are nearly equal in brightness; however, at its minimum, β is only half the brightness of γ. Beta Lyrae is an excellent variable to observe using binoculars. It's bright and so are the comparison stars.

The W UMa (EW) light curve is also continuously variable, similar to the β Lyrae light curve, but with only a small difference in the depths of the minima. The variation outside the eclipse in the latter two types is indeed due to proximity effects, mainly the tidally distorted shapes of the stars, but the EB light curves arise from detached or semidetached binaries, while the EW systems are over-contact.

The expressions *detached, semidetached,* and *over-contact* arise from morphological classification of binaries. Detached systems contain widely separated stars. Semidetached systems are still separated but one star fills its Roche lobe. Contact systems exist when both stars precisely fill their Roche lobes. In over-contact systems, both stars overfill their Roche lobes and establish a common envelope. Such systems can only exist for astronomically significant times if the orbits are circular and the components rotate synchronously.

As an aid to studying the light curves of eclipsing binaries, the deeper minimum in the light curve is called the primary minimum when the difference in depths of the two minima is clearly discerned. The designation may be arbitrary in cases when there is no difference. Astronomers usually compute the decimal fraction of a photometric cycle, called the phase, from the primary minimum. As I'm sure you remember, the phase for other variable stars begins at the *brightest* part of their cycle.

When it's time to designate the primary star, an astronomer's background usually determines how it is done. The definition varies among photometrists, spectroscopists, and theoreticians, and so the designation is not always consistent. Within the context of photometry, the star being eclipsed at primary minimum is usually called the primary star. As you probably understand, this classification is not necessarily one of size or mass but rather related to temperature. For circular orbit binaries, it is the star possessing the greatest brightness per unit area that is eclipsed at primary minimum. In many cases, this star is usually the more massive of the two.

During spectroscopic study, the usage is occasionally confusing. When studying spectral features, the star with the stronger spectral lines, usually the one with the apparently greater luminosity, is most often classified as the primary star. In radial velocity investigations, the primary star is the one with the smaller radial velocity amplitude, which obviously is the more massive star. While the more massive star is usually the more luminous and, as a result the hotter star, there are cases when this is not true. When theoretical studies are considered, this classification situation becomes even more confusing. Within considering the stellar evolution of a binary star system, the designation "primary" sometimes refers to the originally more massive star which can become the lower-mass star because of mass transfer. Confusing? It's best to check, very carefully, when reviewing a journal article or book, to be sure which star is which.

The *Jagiellonian University Observatory,* also known as the Cracow Observatory in Poland (***http://www.oa.uj. edu.pl/ktt/rcznk.html***) maintains a card catalog containing the times of minima and other information on approximately 2000 eclipsing binary stars. The data has been collected at the observatory since the early 1920s. You'll also find the *International Supplement (SAC – Supplemento ad Annuario Cracoviense)* containing ephemerides for one year that include 880 stars recognized as eclipsing binaries (of Algol, β Lyrae, or W Ursae Maioris type).

Dan Burton maintains a nice Web site, *Eclipsing Binary Stars,* where you can finding information regarding these stars including some photometry on β Lyrae and 68 Herculis, software and a model for computing light curves. The address for the Web site is (***http://www.physics.sfasu.edu/astro/binstar.html***).

Besides the three well-known eclipsing star systems, a new class of eclipsing system was introduced in 2000 (IBVS 5135). Known as the planetary eclipsing transit, this configuration requires a planet rather than a companion star to cause the eclipse. If you want to detect extra-solar planets, this is the type of binary system they you'll want to observe.

Along with the eclipsing systems themselves, the *GCVS* includes additional classifications based upon the physical characteristics of stars found within binary systems. Additional classifications are also based upon Roche lobes. Table 7.1 lists the *GCVS* classifications.

E (Eclipsing binary systems)

Observation Key

 Mixed stars
 Mixed amplitudes
 Mixed periods
👁 Visual, CCD/PEP

– These are binary systems with orbital planes so close to the observer's line of sight (the inclination i of the orbital plane to the plane orthogonal to the line of sight is close to 90 deg) that both components (or one of them) periodically eclipse each other. Consequently, the observer finds changes of the apparent combined brightness of the system with the period coincident with that of the components' orbital motion. **GCVS**

Table 7.1. Eclipsing variable stars arranged in alphabetical order by designation

Variable type		Designation (and subclasses)
Algol type	**EA**	Algol-type eclipsing system
β Lyr type	**EB**	β Lyrae-type eclipsing system
Planetary eclipsing type	**EP**	Stars showing eclipses of their planets
W UMa type	**EW**	W Ursae Majoris-type eclipsing variables
RS Canum Venaticorum	**RS**	RS Canum Venaticorum-type systems
– additional classification according to the component's physical characteristics		
	GS	one or two giant components
	PN	one component is the nucleus of a planetary nebula
	RS	RS CVn system
	WD	systems with a white dwarf component
	WR	systems with a Wolf–Rayet component
– additional classification based on the degree of filling of inner Roche lobes		
	AR	AR Lac-type detached system
	D	detached systems with components not filling their inner Roche lobes
	DM	detached main sequence system
	DS	detached system with a subgiant
	DW	detached system like W UMa systems
	K	contact system with both components filling the inner critical surfaces
	KE	contact system/early spectral type
	KW	contact system of late spectral type
	SD	semidetached system in which the surface of the less massive component is close to its inner Roche lobe.

Note. The combination of the above three classification systems for eclipsing binaries results in the assignment of multiple classifications for object types. These are separated by a solidus ("/") in the data field. Examples are: E/DM, EA/DS/RS, EB/WR, EW/KW, etc.

This is the catch-all category for eclipsing binary stars. When the characteristics of the light curve are sufficiently ambiguous, the star will generally be placed within this group. If you're looking for an interesting project, this is a good place to start. Obviously, these stars belong to *some* classification type. Your mission, if you decide to accept it, is to find out which ones, develop the data to prove it, and then to provide your findings to the world.

You might begin such a project by examining the *GCVS*, looking for those stars visible from your latitude. Then determine which season they will be in a position to be observed. Remember, the best time to observe a star is when it transits. Eventually, check some research literature, perhaps the *Information Bulletins on Variable Stars* or the *Astrophysical Data Service*. Both of these resources will be described in Chapters 10 and 11. Check the literature to see if anyone else has conducted some research regarding the stars that you've chosen. There is no need to repeat someone else's efforts without good cause and you might find some information that will assist you in your research. After you've taken these basic steps, develop a detailed observing program. Keep good notes, be persistent and patient. You'll be surprised at what you can find.

EA (β Persei stars)

Observation Key

 Mixed stars
 Mixed amplitudes
Mixed periods
Visual, CCD/PEP

– Binaries with spherical or slightly ellipsoidal components. It is possible to specify for their light curves the moments of the beginning and end of the eclipses. Between eclipses the light remains almost constant or varies insignificantly because of reflection effects, slight ellipsoidality of components, or physical variations. Secondary minima may be absent. An extremely wide range of periods is observed, from $0^d.2$ to $> 10^4$ days. Light amplitudes are also quite different and may reach several magnitudes. **GCVS**

Beta Persei stars are also known as Algol-type variables, named for the group's prototype star, are a subgroup of eclipsing binaries defined according to their distinctive light curve shape (*morphology*). Their brightness remains approximately constant between the eclipses, that is, variability due to the ellipticity effect and/or the reflection effect is relatively insignificant. As a result, the exact moments of the beginning and the end of the eclipses can be determined by carefully examining the light curve.

Eclipses can range from very shallow ($0^m.01$) if partial, to very deep (several magnitudes) if total. The two eclipses can be comparable in depth or can be unequal (Figure 7.3). In a few cases the secondary eclipse is too shallow to be measurable, for example when one star is very cool, or absent altogether when a high eccentric orbit is present.

Light curves of this shape are produced by an eclipsing binary in which both stars are nearly spherical or perhaps only very slightly ellipsoidal. One star may even be highly distorted to the point of filling its Roche lobe, provided it contributes relatively little to the system's total light. This is the case for at least half of the known EA variables.

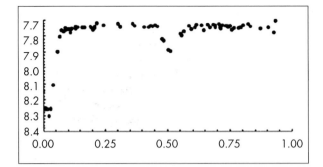

Figure 7.3. Light curve of the EA-type variable star, IQ Per. Data provided by the HIPPARCOS mission. Used with permission.

Among the EAs you'll find binary systems of very different evolutionary status such as:

(a) systems containing two main sequence stars of any spectral type from O to M;

(b) systems in which one or both stars are evolved but have not yet overflowed their Roche lobe;

(c) systems with one star unevolved and the other overflowing its Roche lobe and causing mass transfer;

(d) systems with one star highly evolved, such as a hot subdwarf or a white dwarf, and the other less evolved;

(e) systems with stars not evolved at all.

Binary systems in the third evolutionary state – semidetached, one star evolved and one not, mass transfer in progress – are termed *Algol-type binaries* or *Algol-like binaries*. Such systems, if eclipsing, can have light curves of the EA or EB shape, or they may not eclipse at all. Ironically, β Lyrae, the prototype of the EB light curve shape, is an Algol-type binary.

The first EA discovered, and the prototype of the group, was β Persei. Its variability was known by the Chinese 2,000 years before John Goodricke in 1783 determined the strict period of it variability ($2^d.867$) and first proposed eclipses as the mechanism. Algol has partial eclipses, is semidetached, undergoes mass transfer, has a chromospherically active secondary star which emits radio waves and X-rays, and belongs to a triple system.

For these systems, orbital periods range from extremely short such as a fraction of a day to very long, for example 27 years for ε Aurigae. For an EA light curve shape, the stellar radius or radii must be a relatively small fraction of the star-to-star separation. Note that the brighter component in ε Aurigae is a supergiant but its radius is still a small fraction of the large, 6000 solar radii, semimajor axis.

Orbital periods of EAs can be determined very accurately by timing the sharp eclipses. Period variations are found in many systems. Physical mechanisms responsible can be apsidal motion, orbit around a third body, mass loss and/or mass transfer, and solar-type magnetic cycles. The orbital period of Algol itself undergoes a 1.783 year cycle as it orbits around Algol C and it also has a 32-year magnetic cycle.

EB (β Lyrae stars)

*– These are eclipsing systems having ellipsoidal components and light curves for which it is impossible to specify the exact times of onset and end of eclipses because of a continuous change of a system's apparent combined brightness between eclipses; secondary minimum is observed in all cases, its depth usually being considerably smaller than that of the primary minimum; periods are mainly longer than 1 day. The components generally belong to early spectral types (B–A). Light amplitudes are usually less than 2^m in V. **GCVS***

The β Lyrae stars (EB) are another subgroup of eclipsing binaries segregated according to light curve shape (Figure 7.4). The light curve varies continuously between eclipses, making it difficult to specify the moments of the beginning and the end of the eclipses. To distinguish between EBs and EWs, according to the *GCVS*, the former generally have primary and secondary eclipses significantly different in depth, orbital periods longer than a day, and spectral types B or A.

Light curves of this shape are supposed to be produced by an eclipsing binary in which one or both components is highly ellipsoidal. One of the components may even fill its Roche lobe.

Among the EBs one may find binaries of very different evolutionary status:

(a) unevolved binaries consisting of two main sequence stars but a relatively short orbital period, with XY UMa an example;

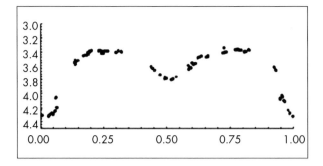

Figure 7.4. Light curve of the EB-type variable star, β Lyr. Data provided by the HIPPARCOS mission. Used with permission.

(b) binaries in which one or both components is evolved but not yet filling the Roche lob, with ζ And an example;

(c) semidetached binaries undergoing mass transfer from the evolved to the unevolved star, with β Lyr an example;

(d) binaries with one star highly evolved, a hot subdwarf or a white dwarf, and the other, producing the ellipticity effects, with AP Psc an example.

Ironically, some binaries classified as EB are not eclipsing at all. The light variation is produced entirely by the ellipticity effect and the two minima are unequal as a result of greater limb-darkening effects on the pointed end of the highly distorted star.

The first EB discovered, and the prototype of the group, was β Lyrae. The same John Goodricke of Algol fame discovered the variability of β Lyr one year later, in 1784. β Lyr is extremely complex and interesting. The brighter star fills its Roche lobe and is transferring matter onto the other star so rapidly that a thick (both optically and geometrically) disk has built up which almost completely obscures the underlying mass-gaining star itself. This mass transfer causes the orbital period to increase at a furious rate. In the 210 years since Goodricke's 1784 timing, the period has increased from $12^d.8925$ to $12^d.93854$, an increase of 0.35%.

Observing β Lyrae stars can be an exciting project for visual observers and many EB variables fall within the visual observing capabilities of amateur astronomers. For those wishing to embrace a deeper challenge, hundreds of EB variables exist that must be examined using photometric methods. In either case, observing these interesting stars will provide a lifetime of enjoyment.

EP (planetary transit eclipsing systems)

– These are stars showing variability as a result of eclipses caused by transits of orbiting planets. On July 9, 2001, The 76th Name-List of Variable Stars (IBVS 5135) added this new classification to the definition used by the General Catalog of Variable Stars, 4th Ed. The prototype star for this class of variable is designated V376 Pegasi (HD 209458). **76th NL**

"Are there somewhere other worlds like ours?" is a question usually ascribed to Epicurus around 300 BC. His school in Athens, the Garden, competed with Plato's Academy and Aristotle's Lyceum. The question can certainly be considered one of humankind's oldest, enduring questions.

In consideration of this question, several detection methods are presently used by astronomers to search for extra-solar planets. One method is well suited to the amateur astronomer and is of singular interest to variable star observers. The transit method – searching for stars that vary in brightness because of a planetary transit – is within the ability of many amateur astronomers. In fact, *amateurs have already detected planetary transits of stars.*

In regards to the transit method, late dwarf M-type stars (dM) are the most attractive candidates for extra-solar planet transit projects because of their small surface areas and lower luminosity when compared to hotter stars. A planet passing in front of one of these cool, small stars will block a larger proportion of the star's total light, producing a greater variation in brightness, than when passing in front of a hotter, larger, brighter star. It can be shown that a planet with a radius that is approximately three times that of the Earth, passing in front of a late M-type star, will produce a 0^m01 drop in brightness. A hundredth of a magnitude seems small but it is detectable using a modest sized telescope equipped with a CCD or by photoelectric methods, both available to amateur astronomers. The light curve of V376, created by a planetary transit, shows an amplitude of approximately 0^m02 (Figure 7.5). There is no question that this level of accuracy is within the reach of amateur astronomers. For CCD users, this means making superior calibration images (i.e. flat frames and dark frames), using long integration times (filters makes this easier when

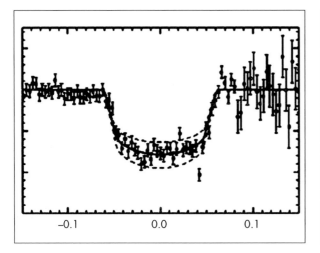

Figure 7.5. Light curve of the EP-type variable star, V376 Peg. Data provided by the High Altitude Observatory/ National Center for Atmospheric Research. Used with permission.

observing bright stars), and by taking steps to minimize the effects of atmospheric extinction and turbulence. Attaining this level of accuracy should not be considered trivial and requires the faithful application of photometric methods and techniques considered by many to be arduous at best. If you want easy, go look at the Moon.

Virtually all past search projects have focused on stars of spectral type G and K. A few surveys looking for planetary transits have been done for stars as late as M6 type; however, no surveys have focused exclusively on stars of spectral type M5 and later. As a result, planetary companions have only been found for stars as late as spectral type M4. As stated earlier, in regard to conducting searches for planetary transits, small surface areas and reduced luminosity make late dwarf stars excellent candidates for observation

It has been understood for some time that the presence of third bodies can disturb the periodicity in the arrival times of planetary transits. As a point of interest, recent contributions have also shown the transit method to be capable of detecting planetary rings and massive planetary moons. These bodies may be detected through deviations from the normal transit shape, or by deviations from the strict periodicity of a transit. An opaque planetary ring, for example, may produce a symmetric transit with a step-wise ingress and egress visible within the light curve, whereas a moon large enough to cause an observable contribution to the light loss in a transit would cause stepped, but asymmetric transit shapes. Furthermore, a moon, even if undetectable in the transit shapes, will cause the associated planet to be either leading or trailing the planet–moon barycenter, thereby causing deviations in the time of the transit from the strict periodicity with the planet's orbital period. In the Earth–Moon system, for example, the barycenter is offset from the geometrical center of the Earth by 4660 km, causing the Earth's center to lead or trail the barycenter by up to 2.6 minutes on its path around the Sun. Similarly, a system consisting of Saturn and its heaviest moon Titan would cause shifts in transit times up to 30.5 seconds. Such time differences would be easily recognizable, once at least three transits have been observed.

Eclipsing binary systems also provide astronomers with a special opportunity to employ the planetary transit method. As shown earlier in this chapter, the inclination of the plane of the eclipsing binary systems is close to 90°. This inclination can be measured precisely from an analysis of the system's light curve. Furthermore, a planetary system is expected to have been precessionally dampened into the plane of the binary components during its formation. For suitable eclipsing binary systems, the probability that planets will cause observable transits is close to 100%. A further advantage of the observation of binary stars is the unique, quasi-periodic transit signals produced therein. Since the system is composed of a double star, there will normally be two transits. The exact shape of the light curve depends on the phase of the binary system at the time of the planetary transit.

A model to consider is the eclipsing binary system CM Draconis that has been monitored with high precision for many years. Periodic deviations of the minimum times may indicate the presence of an orbiting third body. An excellent example of the small, cool dwarf system mentioned earlier, CM Dra is the lowest mass eclipsing binary system known, with components of spectral class dM4.5/dM4.5. The combined surface area of the two stars is about 12% of the Sun's, and the transits of a planet 3.2 time the radius of Earth would cause a drop in brightness of $0^{\text{m}}01$, well within the reach of current differential photometric techniques. CM Dra is relatively close (17.6 pc) and has a near edge-on inclination of 89.82°.

Another method used to search for planetary transit eclipsing systems is called the "search for Trojan extra-solar planets." Consider the Trojan asteroids found in Jupiter's orbit. They are located at the L4 and L5

Lagrangian points. These positions are located 60 degrees ahead of and behind Jupiter and they exist because of the complex gravitational relationships between the Sun, Jupiter and the asteroids.

Using the Trojan planet search method, replace our Sun with a massive star, Jupiter with a low-mass companion star, and the possibility exists of finding planets positioned in the L4 and L5 Lagrangian points relative to the low-mass companion star. The advantage of searching for eclipses using this system is that the timing of the planetary eclipses can be predicted. They are 60 degrees ahead of and behind the stellar eclipse. In eclipsing binary methodology, when we plot the light curve versus phase, from zero to one with zero at mid-primary, this puts the Trojan eclipse of the hotter star centered at phases 0.167 and 0.833.

When using this method, one strategy is to choose targets in which the binary system produces a deep primary eclipse. A deep eclipse usually suggests that one star is significantly hotter (brighter) than the other. Deep eclipses can also indicate nearly edge-on systems. The planetary eclipse will be most evident when observed in a passband that is bluer than the cool star's peak color.

Regarding the observation of EP-type variable stars, current study seems to indicate the need for photometric accuracy equal to or better than $0^{m}01$ when conducting searches for planetary transits. This level of accuracy is well within the reach of serious, dedicated amateur astronomers. With little room for serious argument, this domain of variable star observing can certainly be considered, today, on the "cutting edge" of amateur research.

EW (W Ursae Majoris eclipsing systems)

– These are eclipsers with periods shorter than 1 day, consisting of ellipsoidal components almost in contact and having light curves for which it is impossible to specify the exact times of onset and end of eclipses. The depths of the primary and secondary minima are almost equal or differ insignificantly. Light amplitudes are usually < 0.8 mag in V. The components generally belong to spectral types F–G and later. GCVS

Observation Key	
Mixed stars	
Small amplitudes	
Mixed periods	
⊚ CCD or PEP	

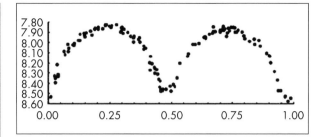

Figure 7.6. Light curve of the EW-type variable star, W UMar. Data provided by the HIPPARCOS mission. Used with permission.

The W Ursae Majoris eclipsing binary stars are characterized by continuous light changes due to eclipses and because of their changing aspects resulting from tidal distortions. The minima in the light curves are of almost equal depth, indicating similar surface temperatures of the components and the periods are short, almost exclusively ranging from about seven hours up to one day (Figure 7.6).

The W UMa phenomenon is usually explained by assuming that both stars are in contact and the more massive star is transferring material to the less massive one through a common envelope. The result is probably an equalizing of the surface temperatures.

Period changes are observed in all EW systems and are probably associated with the ongoing mass circulation that transports the material from the primary to the secondary. The long-term evolutionary effects should produce a secular mass loss of the secondary star resulting in a lengthening of the period, if no matter is lost from the system.

EW systems show complex behavior and period changes. Studies indicate that positive and negative period jumps are randomly distributed among the EW binaries but a comprehensive explanation is lacking. The space distributions of W UMa systems indicates that they form within the old disk population and have a typical age of one billion years. It is suggested that they descend from the short-period main sequence RS CVn systems and then evolve into blue stragglers or into FK Comae-type variables.

Discovered in 1888, S Ant was the first W UMa system. The *GCVS* lists a little over 500 EW systems. Light curves are available for only a small fraction of these stars.

RS CVn (RS Canum Venaticorum eclipsing systems)

– A significant property of these systems is the presence in their spectra of strong CA II, H and K emission lines of variable intensity, indicating increased chromo-spheric activity of the solar type. These systems are also characterized by the presence of radio and X-ray emission. Some have light curves that exhibit quasi-sine waves outside eclipses, with amplitudes and positions changing slowly with time. The presence of this wave (often called a distortion wave) is explained by differential rotation of the star, its surface being covered with groups of spots; the period of the rotation of a spot group is usually close to the period of orbital motion (period of eclipses) but still differs from it, which is the reason for the slow change (migration) of the phases of the distortion wave minimum and maximum in the mean light curve. The variability of the wave's amplitude (which may be up to 0.2 mag in V) is explained by the existence of a long-period stellar activity cycle similar to the 11-year solar activity cycle, during which the number and total area of spots on the star's surface vary. GCVS

Observation Key	
★	Mixed stars
⬚	Small amplitudes
◒	Mixed periods
◉	CCD or PEP

As stated earlier, the classification RS appears in the *GCVS* twice. You saw it earlier as one type of eruptive variable star. Here, the RS Canum Venaticorum system is defined as consisting of binaries in which the hotter of the two is F or G. Other distinctions are made but for our purposes, this will suffice. Other distinctions, not usually considered within the definition, are that they generally have at least one star evolved off the main sequence not yet filling its Roche lobe, they emit intense coronal X-ray and radio radiation, have strong emission lines in the far ultraviolet, lose mass in an enhanced wind, have variable orbital periods, show a starspot wave, and undergo more gradual changes in the mean brightness.

The starspot wave, usually the principal cause of variation within RS variables, is usually sinusoidally shaped.

Discovered in the 1930s, the first non-eclipsing RS CVn binary system that varied in brightness as a result of only starspots was γ Andromedae. The *GCVS* did not officially classify it as an RS-type star until 1985.

Chapter 8

Optically Variable Close Binary Sources of Strong, Variable X-Ray Radiation (X-Ray Sources)

Close binary systems that are sources of strong, variable X-ray emission and which do not belong to or are not yet attributed to any of the other types of variable stars. One of the components of the system is a hot compact object (white dwarf, neutron star, or possibly a black hole). X-ray emission originates from the infall of matter onto the compact object or onto an accretion disk surrounding the compact object. In turn, the X-ray emission is incident upon the atmosphere of the cooler companion of the compact object and is reradiated in the form of optical high-temperature radiation (reflection effect), thus making the area of the cooler companion's surface an earlier spectral type. These effects lead to quite a peculiar complex character of optical variability in such systems.

GCVS

X-ray variable stars! Surely, you jest?

In fact, many X-ray variable stars have an optical component that is within the observational reach of visual observers. Without a doubt, you will fail to detect the X-ray portion of these star's electromagnetic energy spectrum but that's not what you want anyway. And if you're not new to variable star observing, you have probably already viewed stars similar to X-ray variables. Both *low-mass X-ray binaries* (LMXB) and *cataclysmic*

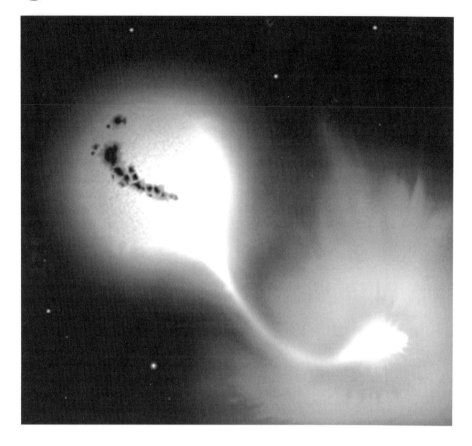

Figure 8.1. Artist's conception of an X-ray-type variable showing the high-energy produced by the accreting material.
Copyright: Gerry A. Good.

variables (CV) have similar orbital periods, low-mass donor stars, mass-transfer rates, variability, and eruptions. The essential difference between CVs and LMXBs is that the accretion is onto a white dwarf for cataclysmic variables and in the case of low-mass X-ray binaries, it is onto a more compact neutron star, or possibly a black hole. In a sense, X-ray variables may provide you with an opportunity to study, in an indirect manner of course, neutron stars and perhaps black holes (Figure 8.1).

With all of that said, I must also caution that as a group this class of variable stars is not a good starting point for beginning amateur astronomers. The great majority of these stars are fainter than 14^m0 and many are as faint as 18^m0 and 19^m0 and those displaying a reasonable brightness are variable for reasons other than X-ray radiation, such as being γ Cas, novae, or emission type variable stars. As you will discover, many variable stars found within the other five major classes of variable stars are also X-ray sources. Still, the intrigue regarding X-ray binaries exists and their study is not beyond the capabilities of serious amateur astronomers. A few years ago who would have believed you could detect planetary transits with a telescope in your backyard? Tough to do, yes! Impossible, no!

X-ray radiation is electromagnetic radiation, similar to visible light but possessing much higher energies. As a result, X-rays exist in the upper energy levels of the spectrum, far beyond the visible, and can only be detected by special X-ray detectors that are usually placed in an orbiting satellite. X-rays find it difficult to penetrate the Earth's atmosphere because they are absorbed by the atmosphere. They indicate high temperatures; very high temperatures and high temperatures reveal high energies. Anything producing a lot of energy, far in excess of what astrophysicists consider normal, is going to be interesting because it's relatively rare or because it will provide an opportunity to observe something that cannot be produced in a laboratory. It's tough to sustain temperatures in the neighborhood of ten million degrees or to produce gravity fields millions of times greater than the Earth's within even the best equipped research facilities on Earth. Perhaps one day.

With all of that said, you can continue your exploration of the Universe through the observation and study of these exotic stars with a little preparation, a lot of patience and a pinch of luck. You won't need your X-ray goggles either. However, there exists several types of X-ray variable stars so you have more decisions to make. The *GCVS* defines the various types of X-ray variables as:

XB – X-ray bursters are close binary systems showing X-ray and optical bursts, their duration being from several seconds to ten minutes, with amplitudes of about $0^{m}\!.1$ in V.

Examples of XB-type variables are V801 Ara and V926 Sco. The star V801 Ara shines at $16^{m}\!.0$ when brightest and V926 Sco at $17^{m}\!.4$. As you can see, these are pretty faint stars and it will require a large telescope to observe stars like this. Fortunately, there is a place that you can visit to better understand X-ray variables when you can't actually observe them yourself. It's the *High Energy Astrophysics Science Archive Research Center* (HEASARC) (***http://heasarc.gsfc.nasa.gov/***).

You can find gamma-ray, X-ray, and extreme ultraviolet observations of cosmic, non-solar system sources at the HEASARC Web site along with archival data, associated analysis software, documentation and guidance in how to use it all, as well as educational and outreach material. This site also provides astronomical tools so that you can obtain multiwaveband images of the sky and conduct astronomical catalog searches. You can also study images, spectra, and light curves from celestial high-energy sources including cataclysmic variables, X-ray binaries, supernova remnants, pulsars and gamma-ray bursts. Another must-visit site!

Now, let's finish looking at the types of X-ray variables.

XF – Fluctuating X-ray systems showing rapid variations of X-ray and optical radiation on time-scales of dozens of milliseconds.

Because of the rapid variations, large telescopes with instruments are required to collect valuable data on fluctuating X-ray systems.

XI – X-ray irregulars are close binary systems consisting of a hot compact object surrounded by an accretion disk and a dA–dM-type dwarf. These display irregular light changes on time scales of minutes and hours, and amplitudes of about 1 magnitude in V. Superposition of a periodic variation because of orbital motion is possible.

The star V818 Sco may give you a good opportunity to observe this type of X-ray variable. Its brightness is about 12^m0, perhaps within the reach of good quality binoculars.

XJ – X-ray binaries characterized by the presence of relativistic jets evident at X-ray and radio wavelengths, as well as in the optical spectrum in the form of emission components showing periodic displacements with relativistic velocities.

The prototype star for this group of X-ray variables is V1343 Aql. This star is also an eclipsing binary with a period of 13.0848 days. At its brightest, it is 13^m0, within the capabilities of medium-sized telescopes.

XND – X-ray, novalike (transient) systems containing, along with a hot compact object, a dwarf or subgiant of G–M spectral type. These systems occasionally rapidly increase in brightness by 4–9 magnitudes in V simultaneously with the X-ray range, with no envelope ejected. The duration of the outburst may be up to several months.

V616 Mon is the prototype star of this group of X-ray variables and is also an elliptical variable. At its brightest, this star is within the range of a good pair of binoculars at $11^{m}\!.26$ but it fades to $20^{m}\!.2$.

XNG – X-ray, novalike (transient) systems with an early-type supergiant or giant primary component and a hot compact object as a companion. Following the main component's outburst, the material ejected by it falls onto the compact object and causes, with a significant delay, the appearance of X-rays. The amplitudes are about 1–2 magnitudes in V.

An O9.7IIIe-type star, the prototype for this type of system is V725 Tau and shines at a bright $9^{m}\!.4$ at its brightest. It will fade to $10^{m}\!.1$, still within the capabilities of a good pair of binoculars.

XP – X-ray pulsar systems. The primary component is usually an ellipsoidal early-type supergiant. The reflection effect is very small and light variability is mainly caused by the ellipsoidal primary component's rotation. Periods of light changes are between 1 and 10 days; the period of the pulsar in the system is from 1 s to 100 min. Light amplitudes usually do not exceed several tenths of a magnitude.

GP Vel, the prototype star for this type of system, is also known as Vela X-1. This system, listed as a B0.5Iaeq-type star, shines within a range between $6^{m}\!.76$ and $6^{m}\!.99$, well within the range of binoculars and small telescopes.

XPR – X-ray pulsar systems featuring the presence of the reflection effect. They consist of a dB–dF-type primary and an X-ray pulsar, which may also be an optical pulsar. The mean light of the system is brightest when the primary component is irradiated by X-rays; it is faintest during a low state of X-ray source. The total light amplitude may reach 2–3 magnitudes in V.

The prototype star for this group of X-ray variables is HZ Her, a B0Ve–F5e-type system. As you might suspect, it is an eclipsing binary pair ranging in brightness from $12^{m}\!.8$ to $15^{m}\!.2$ with a period of $1^{d}\!.700175$. All of the X-ray variables consist of a binary pair so there is a non-zero probability that any particular system is an eclipsing binary.

XPRM – X-ray systems consisting of a late-type dwarf (dK–dM) and a pulsar with a strong magnetic field.

Matter accretion on the compact object's magnetic poles is accompanied by the appearance of variable linear and circular polarization; hence, these systems are sometimes known as "polars." The amplitudes of the light changes are usually about 1 magnitude in V but, provided that the primary component is irradiated by X-rays, the mean brightness of a system may increase by 3 magnitudes in V. The total light amplitude may reach 4–5 magnitudes in V.

Two stars, AM Her and AN UMa, are listed within the *GCVS* as examples of this type of system. You probably recognize AM Her from the chapter on cataclysmic variables. It's listed there as a novalike cataclysmic variable and here it's listed as an XPRM. The second example, AN UMa, is a faint $15^{m}.0$ star, at its brightest.

Now is a good time to point out something but I don't want it to be taken in the wrong context. Within the last few chapters, we've been examining the various classes of variable stars; their characteristics, how they're classified, where we can find catalog data and how to observe them. As I stated earlier, the classification of variable stars cannot be taken lightly. It's difficult and the compilation of the various catalogs and databases requires an enormous effort by dedicated people. It's inevitable that some ambiguous characteristics will cause you some confusion.

The cataclysmic variable star AM Her has just been identified as an example of the XPRM type of X-ray binary system. We also know that this star is recognized as a cataclysmic variable, specifically a novalike subclass known as a "polar." As you can see, a little confusion is beginning to emerge here. If you were to check the SIMBAD database, a great resource with a huge amount of information regarding stars, you will find that AM Her is listed as a δ Scuti type variable, spectral type M4.5, and AN UMa is listed as a cataclysmic variable with a spectral type of CV (at least, at the time of the writing of this book).

Obviously, something is wrong but you should be able to recognize the problem without much effort. Here is an example of both classification ambiguity and of human error; nothing spectacular and it certainly doesn't need to become a subject of profound confusion but it is typical of the type of puzzlement that you will find. When detected, errors should be reported so that the problem can be remedied.

Let us apply a little logic and sort this problem out. In this case, we know that AM Her stars and all cataclysmic variables in no way resemble fast pulsating stars such as δ Scuti variables other than that they are all stars. Regardless of the differences in the underlying physics, simply looking at a light curve will bear this out. Also, the spectral type listed for AM Her within the SIMBAD database, M4.5, is well outside of the boundary in which we find the δ Scuti stars. If you look carefully at the description of the XPRM stars you will see that they consist of *a late-type dwarf (dK–dM) and a pulsar.* Since we know that accretion disks, such as are found around these types of stars, cause a *peculiar* spectral type (pec) and that these systems are binary, the simple answer is that the spectral type should be listed as pec+M4.5 and that the δ Scuti classification is simply a human error. In the case of AN UMa and the SIMBAD database, human error is responsible too. Again, all of this should be apparent when you see it and it shouldn't really cause you much of a problem. What *will* cause you problems is when an error is more subtle, for example when a star is misclassified as a variable that exhibits the characteristics of several unrelated classes or when the spectral type is slightly in error. Certainly you would have no problem detecting an error if a Mira type variable was listed as having a spectral type of B9V.

The point of all this? Errors exist within the catalogs and databases that you are going to use. Sometimes the error will be small and sometimes it will be large. When you find errors, do a little research and you'll work yourself clear of the confusion. Apply a little logic, don't jump to conclusions and enjoy the intellectual challenge. Don't get annoyed either. The people providing all of these services, catalogs and databases, are doing their best. Just let them know when you find an error.

I've placed *gamma ray bursts* (GRBs) within the X-ray binaries in this book because of their behavior and the amount of energy they produce. When you first think about GRBs, you may consider the classification of eruptive variables but we really don't see violent processes and flares occurring within a star's chromosphere and corona. In fact, we don't even see a star that can be examined for very long. You might even consider cataclysmic variables but the energy produced within a GRB flash far exceeds the energy produced from accretion material surrounding, or impacting, a white dwarf or even the collapse of a Type II supernova. The energy produced within GRBs indicates that X-ray binary systems is a good place for them to reside right

Table 8.1.

Time after burst	Maximum visual magnitude	Minimum visual magnitude
10 min	12^m6	15^m6
30 min	14^m0	16^m6
1 hour	14^m9	17^m4
2 hours	15^m8	18^m5
4 hours	16^m6	19^m7
6 hours	17^m2	20^m3
24 hours	18^m2	24^m0

now. Perhaps, it may even be that GRBs end up being a new class of variable stars.

Astronomers are struggling to provide a good explanation for these enigmatic bursts of energy. When a GRB is detected, no star is known to be at the position of the burst and the burst fades quickly, sometimes within minutes. After the burst has faded, a check of the area again fails to uncover a star. To give you an example of how fast these events appear and then fade, Table 8.1, provided by Scott Barthelmy (NASA-GSFC) and Jerry Fishman (NASA-MSFC), shows how the brightness of a typical faint GRB afterglow might be expected to diminish with time. Fading fireballs can be fainter than 20^m0 just a few hours after the onset of the explosion.

We are not going to spend a lot of time on X-ray variables. I've mentioned them here so as to be as complete as possible. They cannot be recommended for beginning variable star observers nor can they be adequately presented in this book. In almost all cases, instruments such as CCD cameras or photoelectric devices such as photometers must be used to adequately study these stars.

Chapter 9
A General Approach to Observing Variable Stars

The amateur astronomer has access at all times to the original objects of his study; the masterpieces of the heavens belong to him as much as to the great observatories of the world. And there is no privilege like that of being allowed to stand in the presence of the original.

Robert Burnham Jr.

"Is observing variable stars different from observing other stars, galaxies, nebulae, planets or the Moon?" poses a common concern for beginning variable-star observers.

The answer to your question is "Yes!"

The unique obligation of variable-star observing (VSO) is comparing the variable star with a non-variable star, the comparison star, in order to estimate its brightness. In no other astronomical endeavor is this the principal objective. An accurate estimate of the brightness of the variable star is critical; it is after all, the name of the game. Because the estimate is crucial, much should be considered. We'll advance from general concerns to specific concerns over the next few chapters.

Using Your Eyes to Observe

First, let's take a few moments to learn about how your eyes work. Your ability to effectively use your eyes will be of primary importance when you begin observing variable stars because you are going to push them to the limits of their capabilities. A little understanding of how they work will go a long way in helping you get the most out of your time spent observing, regardless of whether you are using binoculars, a telescope or just your unaided eyes.

Once you understand how your eyes work, you will have several choices regarding how you want to make your observations. We will start with visual observations: those made using only your eyes and perhaps some type of optical arrangement such as binoculars or a telescope. Instrument observations are made using CCDs or photoelectric devices, in which case you may not even "look" at the variable star that you are studying; instead the instrument measures and compares the brightness of the stars.

As light detectors, your eyes are exceptionally adaptable organs, able to produce remarkable results in both bright sunlight and faint starlight. Consisting of four important elements, your eyes are able to convert light into an electrical signal that is transmitted to the brain where it is interpreted. These four important elements of your eye are: the cornea, the transparent part of your eye; the lens, providing the focusing ability; the iris, the pigmented diaphragm that restricts the aperture and the amount of light allowed to enter your eye; and the retina, a lining of nerve cells along the back of your eye that detects light and then converts it into nerve impulses.

When light strikes the nerve cells within your retina, a photochemical reaction produces a nerve impulse that sends an electrical signal to your brain. It is within the nerve cells found lining your retina that light is detected. As an astronomer, the ability of your eyes to detect light is critical since that is your ultimate goal. Detecting light and the sensitivity of your eyes is controlled by chemistry. Essentially, by varying the amount of photosensitizing chemicals within your eyes, *not by opening your iris*, they can adapt to a wide range of intensity levels.

These light-detection nerve cells within your retina are called *rod cells* and *cone cells*. The cone cells are concentrated in only a small, central region within each eye. This small area, called the *fovea*, defines the center of your vision. Normally, you aim this region directly at an object to see the most detail.

These centrally located cone cells are not as sensitive to light as the peripheral rod cells; however, there are no rods in the center of the fovea. Instead, the rods increase in density from zero at the center of the fovea to a maximum at about 18 degrees out from the center of the fovea.

There are no rods or cones on the spot where the optic nerve leaves the eye. This is the blind spot and no vision is possible from this spot. To help you compensate, your eye tends to jump around when you are looking at something, partly to ensure that the blind spot does not hide something important. You can find your blind spot by simply raising a hand in front of your face, about a foot (30 cm) out and at eye level, and form a "V" with your first two fingers. Close one eye and stare at the fingernail of your inside finger (*if you are looking down your right arm with your right eye, stare at the left most finger*) while slowly moving your hand away (with a slight adjustment up or down) from your face. You will notice that you can see the fingernail of your outside finger when you first begin. Eventually, at a certain distance out from your face, you will notice that the fingernail of your outside finger disappears. You will probably notice that the whole top of your finger disappears. This happens because the light entering your eye from that angle hits the blind spot. There are no light-sensitive cells to detect the light. With a little practice, while looking out of a window, you can cause bushes, trees and small buildings to disappear using this technique. At the eyepiece, you can cause an entire star to vanish!

One of the interesting behaviors of your eyes, important for variable-star observers, is that the rod cells and cone cells each respond best to different colors of light. On average, the rod cells are slightly more blue-sensitive than the cone cells. However, it's the cone cells that are entirely responsible for your color vision. Cone cells require bright light to work best and that is why you do not see color in dim light. In dim light, only the rod cells are working efficiently.

Dark Adaptation and VSO

When exposed to dim light, the pupil of your eye can open within a couple of seconds to about 7 or 8 millimeters. As you have no doubt noticed, you cannot see very well in dim light. When your iris opens in response to dim light, the amount of light entering your eye increases by no more than a factor approaching sixteen. However, you have also noticed your eye's sensitivity increases, as times goes by, by a factor of many thousands. This process is called *dark adaptation*.

Dark adaptation is the result of a chemical process. In dim light, a chemical called *rhodopsin* is produced within your eye and it concentrates within the rod and cone cells. The quantity of rhodopsin determines the sensitivity of your eyes. Dark adaptation that is adequate for variable-star observing is usually complete after about ten or fifteen minutes. Longer time is needed for observing very faint stars. You can begin this dark adaptation process early in the evening by wearing an eye patch an hour or so before you begin to observe.

Another characteristic of your eye that is important to you as a variable star observer is *sensitivity*. The sensitivity of your eye will ultimately determine how accurately you will be able to compare the brightness of stars. As you know, comparing the brightness of stars is the critical element of variable-star observing.

You have probably noticed that when the apparent size of an object increases you are able to better see it, regardless of dark adaptation. In other words, the bigger something is, the less need you have for dark adaptation. If you live out in the country and step out your back door, you will be able to see a barn, since it's so large, but you'll have difficulty in locating a door from a distance (unless you move closer and the door appears larger). As your eyes adapt, fine details, such as doors and fences become visible. *This is a critical concept for you to understand.* You will find that not only the brightness but the size of an object as seen in the eyepiece will affect its visibility. The importance of this will become apparent in a moment.

Variable-star observation depends not just on detecting a faint star but also on contrast discrimination. For stars that are just detectable, you will notice that as the background sky becomes darker, the size of the star appears larger. Your eye's ability to see a point source, such as a star, increases as the background glow decreases, in other words, as the sky darkens. Dark country skies are better than city skies for seeing dim stars because the background is less bright, not entirely because the country skies are significantly more transparent.

If an object is at your eye's threshold of detection and smaller than the optimum size for your eye and light conditions, increased magnification will usually make it easier to see. By increasing the magnification you decrease the brightness of the background sky and improve your ability to see a faint star. This phenomenon is the reason that you will need more than one or two eyepieces for observing variable stars. A selection of eyepieces will allow you to configure an optical path best suited for existing light conditions, your eyes and the brightness of the stars that you are observing. Of course, you can go too far and over-magnify an object. It's advisable to slowly increase the magnification until the star that you are observing is bright enough to estimate. If you over-magnify, the star will become fuzzy and start to fade. If this happens, just go back to the last eyepiece that provided the best view. Remember, as you increase magnification you decrease field-of-view (FOV). That means that your comparison star may not be visible within your FOV. It's a good idea to record in your log or record book, the eyepiece that you determine to be best for a particular star so that you can quickly select the appropriate one during observing. It saves time.

Averted Vision and VSO

When observing a faint light source and looking off to the side you are using *averted vision*. Most amateur astronomers quickly learn about averted vision and use it to observe faint objects that are not visible when viewed straight-on.

The eye seems to have a limited integration capability similar to photographic film. For the detection of the faintest objects, it seems that light must collect on the retina for several seconds. If you have ever strained at an eyepiece, attempting to view a star just at the visual limit of your equipment until you where able to finally catch a fleeting glimpse of it, then you know this to be true.

When using averted vision it's important to take your time. When searching for faint objects within your FOV, you will miss many faint stars if you scan the area too fast. Concentrating on a point within your FOV while using averted vision will allow you to see many stars that are too faint to observe when viewed directly.

It takes practice to concentrate on a point, usually without a star present, while using averted vision because the eye tends to jerk around slightly (remember the blind spot). You will find that fatigue also compounds the problem. This is why it's a good idea to take frequent, short breaks and to eat something while observing for more than an hour or so. Observing really is a physical effort.

How Color Affects VSO

As you know, the human eye is an extraordinary detector of color under bright conditions. You also know that the color receptors, the cone cells, do not function well at night and as a result you will usually see no color. You have no doubt noticed that you are unable to distinguish the colors of vehicles or of the clothing of people walking in parking lots at night or along a street that is poorly illuminated.

Because rod cells and cone cells are each most sensitive to different colors, the perceived brightness of an object near the transition light level when your eyes are switching from cone cells to rods cells, for example when the sun is setting, can depend on its color. This *Perkinje effect* is known, but not well understood, to variable-star observers who often have to compare two stars of different colors. As you will discover, it's not easy to find a comparison star that is the same color as the variable that you are observing.

The Perkinje effect is named for Jan Evangelista Purkinje, a pioneering Czech experimental physiologist whose studies within the fields of histology, embryology, and pharmacology helped create a modern understanding of the eye and vision. The Perkinje effect describes the following phenomenon: *As light intensity decreases, red objects are perceived to fade faster than blue objects of the same brightness.*

In 1819, Purkinje noticed that as the light faded on his garden the red poppies became black but the blue flowers remained blue and the green leaves stayed green. Purkinje discovered that when light levels decrease, the human eye becomes more sensitive to blue and green light. Purkinje theorized,

The implication is that this twilight zone is the most dangerous one: lighting conditions are varying quickly; nocturnal predators are appearing: and fatigue is setting

in. It might well be more adaptive to have better vision during this brief but dangerous period than to optimize reception to the spectrum of moonlight when the light levels are too low to allow any advantage to be had from it.

Basically, he said that perception is more important than color.

Despite their individual weaknesses, the rod and cone systems make a great team, working together as two hunters who have agreed that one will scan the countryside for movement but must allow the other to identify what it is that's moving. Of course, we do not notice that we have two receptor systems because our brain seamlessly combines their outputs.

All of this will become important when you begin to observe bright red stars, such as the Mira and semiregular type variables or when you must compare a variable with a comparison star of a different color. Later, we will take a look at some of the methods you can use to reduce the Perkinje effect when observing variable stars.

Using Binoculars for VSO

Using binoculars to observe variable stars is an excellent decision because binoculars offer several advantages over a telescope when it comes to observing variable stars. First, they are easy to use and require a minimum amount of effort to set up and take down. Secondly, binoculars offer a wider field of view so that finding comparison stars is relatively easy.

If you believe that binoculars will not allow you to make detailed observations or will in some way inhibit your ability to detect subtle changes in the brightness of stars, take a look at some light curves produced by Kari Tikkanen, an observer from Finland who uses binoculars to observe variable stars. Kari uses 10×50 and 12×50 binoculars and reports his observations to the *Swiss Astronomical Society* (BBSAG) as well as the *American Association of Variable Star Observers* (AAVSO). His timings are published in the BBSAG bulletin.

The first light curve is of the star HU Tau, a semidetached, eclipsing binary of the Algol type (EA/

SD) with a period of 2^d056. At maximum light its brightness is about 5^m85 and at minimum about 6^m68 (Figure 9.1). As you can see, Kari has been able to plot the eclipse very well with 78 observations. As a testament to his accuracy and ability, Kari records his observation to an accuracy of two decimal points (a hundredth of a magnitude).

The second light curve is of the star U Sge, another eclipsing binary. This star has a period of 3^d38, a maximum brightness of about 6^m45 and a minimum of 9^m28 (Figure 9.2).

Using 206 observations, Kari was able to observe the full eclipse (down to 9^m3) with exceptional accuracy. It should be obvious that binoculars will not limit your ability to observe variable stars! Nor will stars with small amplitudes pose any serious hurdle to serious observers. There are many hundreds, perhaps thousands, of stars suitable for observing with suitable binoculars. Let's look at what makes binoculars suitable for variable-star observing.

The sharpness and brilliance of any image that you see when looking through a particular binocular are determined by a number of different factors, including the interaction of lens diameter, magnification, optical coatings and design. Overall, most binocular observers will agree that the most important consideration regarding binocular performance is the quality of the optics.

Magnification is the degree to which the object being viewed is enlarged. For example, with a 10×50 binocular, the number 10 indicates the magnification power. The level of power affects the brightness of an image, so the lower the power of a binocular, the brighter the image. In general, increasing power will reduce both FOV and eye relief (the distance from your

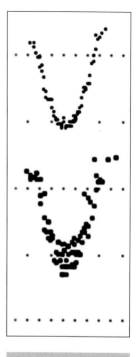

Figure 9.1. Light curve of HU Tau. Data provided by the Kari Tikkanen, URSA VSS. Used with permission.

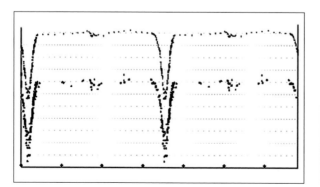

Figure 9.2. Light curve of U Sge. Data provided by the Kari Tikkanen, URSA VSS. Used with permission.

eye to the lens). Remember, if you wear glasses, you can observe without them by using the binoculars as your corrective lens. I wear glasses and contact lenses and I find my vision at the telescope to be noticeably sharper without corrective lenses so I use a strap to keep my glasses close by, for when I need to look at a chart or atlas ... or find my food. When visually observing, I do not wear my contact lenses either.

The objective lenses of binoculars are the front lenses. The diameter of one of these lenses, given in millimeters, will be the second number describing a particular binocular. A 10×50 binocular has an objective lens of 50 mm. The diameter of the lens determines the light-gathering ability of the instrument with the greater light-gathering ability of a larger lens usually translating into greater detail and image clarity. Doubling the size of the objective lenses quadruples the light-gathering ability of the binocular.

The size of the area that can be seen while looking through a pair of binoculars is described as the field-of-view. Field-of-view is related to magnification, with greater magnification creating a smaller field of view. Low-power, therefore a large field-of-view, is especially desirable when observing variable stars because you will see more comparison stars. It's also easier to figure out were you are in the sky with a larger FOV because more stars are visible.

Eye relief refers to the distance, in millimeters, that a binocular can be held from the eye and the full field of view can still be comfortably observed. Eyeglass wearers in particular benefit from longer eye relief.

The optical elements of the binocular are coated to reduce internal light loss and glare, that in turn insures even light transmission, resulting in greater image sharpness and contrast. Choosing a binocular with good lens coatings will translate into greater satisfaction. Lens coatings range in quality as follows: coated, fully coated, multicoated, fully multicoated. Coated lenses are the lowest quality. Fully coated lenses are economical and may work well for you. Multicoated or fully multicoated lenses are both very good choices but the price is higher. Fully multicoated lenses give the best light transmission and brightest images but the price is higher.

A critical factor in the performance of any binocular is its construction. The security of the barrel alignment and proper internal mounting and alignment of the optics are crucial to producing a binocular that's mechanically reliable, smooth functioning and long-lasting. Check this carefully.

The alignment of the optical elements of the binocular to the mechanical axis is called collimation. Good collimation prevents eyestrain, headaches, inferior and double images while improving resolution. Unfortunately, proper collimation is almost impossible to achieve in very low-priced binoculars that lack quality components and design.

As you see, there are a number of different factors to consider in choosing a binocular. Perhaps the first thing to understand is that binoculars are really just two small telescopes mechanically linked together. All of the facts and formulas that help you to understand telescopes also pertain to binoculars. Each side of a pair of binoculars has a prime focal length, an objective lens, an ocular (eyepiece), an exit pupil, and so forth. Even if you plan to observe using binoculars only, the following section pertaining to telescopes is worth your time.

Using a Telescope for VSO

Do not worry about what telescope works best for variable-star observing. Use what you have or what you can afford. If you are new to amateur astronomy, I will tell you something that will take years to discover on your own. You will never be satisfied with the telescope that you have. It will be too small, too big, too heavy, too light, lacking a computer system, without motor drive ... the list is perpetual. You must, at some point, just say "enough is enough," and use what you have to its full potential. A good telescope, regardless of size, will provide you access to more variable stars than you can observe and study within your lifetime.

However, if you haven't reached that point where "enough is enough," and you're thinking about purchasing a new telescope, wishing to specialize in variable-star observing, then this section may provide some ideas that could be valuable. If you already have a telescope, I think that you'll find some value within this section too. The most important thing is to understand how your telescope works. I mean how it *really* works! It's crucial for you to use your telescope's full capability and observing variable stars will require you to use all of the capabilities of your telescope. Your under-standing of a few important terms is essential.

The single most important factor to a variable-star observer is *aperture*. The primary function of all telescopes is to collect light and at any given magnification, the larger the aperture, the better the image will be. The clear aperture of a telescope is the diameter of the objective lens or primary mirror specified in either inches, centimeters (cm) or milli-meters (mm). With all things being equal (this is of course never the case), the larger the aperture, the more light the telescope collects and the brighter and better the image will be for the observer. Greater detail and image clarity will be apparent as aperture increases. Considering your budget and portability requirements, select a telescope with as large an aperture as possible. This is the one physical characteristic of your telescope with which you will never be satisfied. All amateur astronomers suffer from aperture fever and it's chronic. You will always want a bigger telescope. It's not just a "guy" thing either.

Focal length (FL) is the distance in an optical system from the lens, or primary mirror, to the point where the image is considered to be in focus. This point is called the *focal point*. The longer the focal length of the telescope, generally the more magnifying power, the larger the image and the smaller the field of view. For example, a telescope with a focal length of 2000 mm has twice the power and half the field of view of a 1000 mm telescope when using the same eyepiece. Most manu-facturers specify the focal length of their various instruments; but if it is unknown and you know the focal ratio you can use the following formula to calculate the focal length: focal length is the aperture (in mm) times the focal ratio. For example, the focal length of an 8 inch (203 mm) aperture with a focal ratio of $f/10$ would be $203 \times 10 = 2030$ mm. This would normally be rounded to 2000 mm.

Resolution is the ability of a telescope to render detail. The higher the resolution, the finer the detail. With all other things being equal, the larger the aperture of a telescope, the better the resolution. Resolution is not the most important criterion with which to judge a telescope for variable star observing since you are not going to see any detail on any star except the Sun (with proper filters, of course!). What this means is, you don't need to buy a really expensive refractor when a reflector, with a smidgen less resolution, is less expensive. For the same cost, the reflector is usually bigger too.

Light gathering capability (light grasp) is the telescope's theoretical ability to collect light compared to your fully dilated eye. It is directly proportional to the square of the aperture. You can calculate this by first dividing the aperture of the telescope (in mm) by 7 mm (average size of a dilated eye) and then squaring this result. For example, an 8 inch telescope has a light gathering power of 843 (check the math: $(203/7)^2 =$ 843). This means that an 8 inch telescope will collect 843 time more light than your naked eye. It should be obvious that even a small telescope is going to collect much more light than your eye. If you understand this, you also understand that you do not need a huge telescope to observe variable stars.

Magnification is the least important factor and magnification is a relationship between two independent optical systems: the telescope itself, and the eyepiece that you are using.

To determine magnification power, divide the focal length of the telescope (in mm) by the focal length of the eyepiece (in mm). By exchanging an eyepiece of one focal length for another, you can increase or decrease the magnification power of your telescope. For example, a 25 mm eyepiece used on an *f*/10, 8 inch (FL = 2000 mm) telescope will yield a power of 80× (check the math: 2000/25 = 80) and a 12.5 mm eyepiece used on the same instrument would yield a power of 160 × (2000/12.5 = 160). Since eyepieces are interchangeable, a telescope can be used with a variety of powers for different applications by simply changing eyepieces.

There are practical upper and lower limits of magnification for telescopes. These limits are determined by the laws of optics and the nature of the human eye. As a rule of thumb, the maximum usable magnification is approximately 50 times the aperture of the telescope (in inches) under ideal conditions. Magnification higher than this usually results in a dim, low-contrast image. For example, the maximum magnification possible for a 60 mm telescope (2.4 inch aperture) is approximately 140×. As magnification increases beyond a certain point, the sharpness and detail will be diminished. This is why the large magnification advertised for some small telescopes is not possible. I've seen 1200× listed for some small telescopes. This is absolutely impossible!

Anyway, most of your observing will be done with low powers. With lower powers, the images will be much brighter and crisper, providing more enjoyment and satisfaction with the wider fields of view. Another advantage to using lower power is that your greater FOV will make comparison stars easier to find.

There is also a lower limit of magnification, usually between three and four times the aperture of the telescope when using your telescope at night. Magnification lower than these limits is not useful with most telescopes and a dark spot may appear in the center of the eyepiece in a catadioptric or Newtonian telescope because of the secondary or diagonal mirror's shadow.

If you're new to astronomy, you may be wondering how far you can see with a telescope; in other words, can you see far enough to observe a variable star. Your friends will eventually ask this question too. Astronomers use a system of magnitudes to indicate the brightness of a star. A star is said to have a certain numerical magnitude. The larger the magnitude number, the fainter the star. Each star with an increased number (next larger magnitude number) is approximately 2.5 times fainter. The faintest star you can see with your unaided eye (no telescope) is about sixth magnitude (from dark skies) whereas the brightest stars are magnitude zero (or even a negative number). So you see, it's not really a question of "how far" but "how bright."

The magnitude of the faintest star that you can see with a telescope (under excellent seeing conditions) is referred to as the *limiting magnitude* of your telescope. The limiting magnitude is directly related to aperture, so that larger apertures allow you to see fainter stars. A rough formula for calculating visual limiting magnitude is: 7.5 + 5 LOG(aperture in cm). For example, the limiting magnitude of an 8 inch aperture telescope is about 14.0 (check the math: 7.5 + 5 LOG 20.32 = 7.5 + (5 × 1.3) = 14.0). Atmospheric conditions and your visual acuity will often reduce the limiting magnitude a bit; however, it's not unusual to see stars a bit fainter than the limiting magnitude of your telescope on exceptional nights. The more time you spend looking through your telescope, the better you will be able to see faint stars. It takes time.

Using a CCD, you can extend the limiting magnitude by three, or four more magnitudes. With the proper application of colorful language, even another two magnitudes are possible.

Focal ratio is the ratio of the focal length of the telescope to its aperture. To calculate, divide the focal length (in mm) by the aperture (in mm). For example, a telescope with a 2030 mm focal length and an aperture of 8 inches (203 mm) has a focal ratio of 10 (check the math: 2030/203 = 10). This is normally specified as *f*/10.

Some astronomers equate focal ratios with image brightness but strictly speaking this is only true when a telescope is used photographically and then only when taking pictures of so-called *extended objects* like the Moon and nebulae. Telescopes with small focal ratios are sometimes called *fast* and will produce brighter images of extended objects on film and thus require shorter exposure times. Generally speaking, the main advantage of having a fast focal ratio with a telescope used visually is that it will deliver a wider field of view. Generally, *fast* focal ratio telescopes are considered to be *f*/3.5 to *f*/6, *medium* focal ratios are *f*/7 to *f*/11, and *slow* focal ratios are *f*/12 and longer. Of course, these boundaries are soft and some dispute should be expected.

The amount of sky that you can view through a telescope is called the *true field-of-view* and is measured in degrees of arc (angular field). The larger the field of view, the larger the area of the sky you can see. Angular field of view is calculated by dividing the power being used into the apparent field of view, using degrees, of the eyepiece being used. For example, if you were using an eyepiece with a 50 degree apparent field, and the power of the telescope with this eyepiece was 100×, then the field of view would be 0.5 degrees (check the math: 50/100 = 0.5).

Manufacturers will normally specify the apparent field (in degrees) of their eyepiece designs. The larger the apparent field of the eyepiece (in general), the larger the real field of view and thus the more sky you can see. Lower powers used on a telescope allow much wider fields of view than do higher powers. As I said before, you'll use low power more than high power when observing variable stars.

There are several optical designs used for telescopes. Remember that a telescope is designed to collect light and when designing optical systems the optical engineer must make tradeoffs in controlling aberrations to achieve the desired result of the design.

Aberrations are any errors that result in the imperfection of an image. Such errors can result from design or fabrication or both. It is impossible to design an absolutely perfect optical system. The various aberrations due to a particular design are noted in the discussion on types of telescopes. In general, various aberrations that you will discover are discussed next.

Chromatic aberration is usually associated with objective lenses of refractor telescopes. It is the failure of a lens to bring light of different wavelengths (colors) to a common focus. This results in a faint colored halo (usually violet) around bright stars, the planets and the Moon. Achromatic doublets (a lens arrangement) in refractors help reduce this aberration and more expensive, sophisticated designs like apochromats and those using fluorite lenses can virtually eliminate it. Mild chromatic aberration should not effect your ability to observe variable stars.

Spherical aberration causes light rays passing through a lens, or reflected from a mirror, at different distances from the optical center to come to focus at different points on the axis. This causes a star to be seen as a blurred disk rather than a sharp point. Most telescopes are designed to eliminate this aberration but it's best to purchase telescopes with hyperbolic mirrors rather than spherical mirrors (of course, this pertains to reflectors only).

Coma is associated mainly with parabolic reflector telescopes that affect the off-axis images and is more pronounced near the edges of the field of view. The images seen produce a V-shaped appearance. The faster the focal ratio, the more coma that will be seen near the edge although the center of the field (approximately a circle, that is defined (in mm) as the square of the focal ratio) will still be coma-free in well-designed and manufactured instruments.

A lens aberration that elongates your images causing them to change from a horizontal to a vertical position on opposite sides of best focus is called *astigmatism*. It is generally found in poorly made optics or when collimation errors are present.

Field curvature is caused by the light rays not all coming to a sharp focus in the same plane. The center of the field may be sharp and in focus but the edges are out of focus and vice versa.

Collimation is a word used to mean the proper alignment of the optical elements in a telescope and proper collimation is critical for achieving optimum results. Poor collimation will result in optical aberrations and distorted images. The alignment of the optical elements is important but more critical is the alignment of the optics with the mechanical tube. This is called optical/mechanical alignment. Time must be taken to check this occasionally.

Types of Telescopes

Hans Lippershey of Middleburg, Holland, gets credit for inventing the *refractor* telescope in 1608. Following its invention, the various military organizations of Europe began using the telescope. Eventually, with some minor modifications, Galileo was the first to use it in astronomy. Both Lippershey's and Galileo's designs used a combination of convex and concave lenses. In about 1611, Kepler improved the design so as to have two convex lenses, rendering the image upside-down. Kepler's design is still the major design used for refractors today with a few improvements in the lenses and the glass, of course.

Refractors are the type of telescope with which most of us are familiar. They have a long tube, made of metal, plastic, or wood, a glass combination lens at the front end (*objective lens*), and a second glass combination lens (*ocular*). The tube holds the lenses in place at the correct distance from one another. The tube also helps to keeps out dust, moisture and light that would interfere with forming a good image. The objective lens gathers the light, and bends or refracts it to a focus near the back of the tube. The eyepiece brings the image to your eye and magnifies the image. Eyepieces have much shorter focal lengths than objective lenses.

Refractors have good resolution; however, it is difficult to make large objective lenses, greater than 4 inches or 10 centimeters, for refractors. Refractors are relatively expensive, if you consider the cost per unit of aperture.

Sir Isaac Newton developed the *reflector* around 1680 in response to the chromatic aberration problem with refractor telescopes. Replacing the objective lens, Newton employed a curved, metal mirror to collect the light and then reflect it to a focus. Mirrors do not have the chromatic aberration problems found in lenses. In his new reflector design, Newton placed the large primary mirror in the bottom of the tube.

Because the mirror reflected light back into the tube, he had to use a small flat mirror in the focal path of the primary mirror to deflect the image out through the side of the tube to the eyepiece. Because of its small size, compared to the primary mirror, the smaller mirror will not block the image.

In 1722, John Hadley improved the design by using parabolic mirrors. The Newtonian reflector is considered a highly successful design and remains one of the most popular telescope designs in use today.

Rich-field, or *wide-field reflectors,* are a type of Newtonian reflector with short focal ratios and low magnification. These telescopes offer wider fields of view than longer focal ratio telescopes and provide bright, panoramic views. As you now know, this increases the number of comparison stars that are visible.

Dobsonian telescopes are a type of Newtonian reflector with a simple tube and alt-azimuth mounting (mounts are discussed in the next section). They are relatively inexpensive to build or buy. Commercial Dobsonians usually have large apertures ranging in size from 6 inches to 17 inches. You will see very large Dobsonians at star parties, some with apertures in excess of 36 inches (talk about aperture fever!). Dobsonian telescopes are excellent for observing faint variables, for example, when your primary objective is to visually search for cataclysmic variable outbursts or to hunt for supernovae in faint galaxies.

Compound or *catadioptric telescopes* are hybrid telescopes that have a mix of refractor and reflector elements in their design. The first compound telescope was made by German astronomer Bernhard Schmidt in 1930. The *Schmidt* telescope had a primary mirror at the back of the telescope and a glass corrector plate in the front of the telescope to remove spherical aberration. The telescope was used primarily for photography because it had no secondary mirror or eyepieces. Instead, a photographic film was placed at the prime focus of the primary mirror. Today, the *Schmidt–Cassegrain* design, invented in the 1960s, is considered by some to be the most popular type of amateur telescope. It uses a secondary mirror that reflects light through a hole in the primary mirror to an eyepiece. The Hubble Space Telescope is a Schmidt–Cassegrain design.

A second type of compound telescope was invented by a Russian astronomer, D. Maksutov, although a Dutch astronomer, A. Bouwers, came up with a similar design before Maksutov. The *Maksutov telescope* is similar to the Schmidt design, but uses a corrector lens. Today, the *Maksutov–Cassegrain* design is similar to the Schmidt–Cassegrain design.

Telescope Mountings

Your telescope must be supported by a mount. The telescope mount allows you to keep the telescope pointed toward the stars and to adjust its position relative to the movement of the stars caused by the Earth's rotation. It also allows you to free your hands for other activities such as focusing, changing eyepieces, and record keeping.

There are two basic types of telescope mounts: *alt-azimuth*, and *equatorial*. The alt-azimuth mount has two axes of rotation, a horizontal axis and a vertical axis. To point the telescope at an object, you rotate it along the horizon, the azimuth axis, to the object's horizontal position, and then tilt the telescope, along the altitude axis, to the object's vertical position. This type of mount is simple to use.

Although the alt-azimuth mount is simple and easy to use, it does not efficiently track the motion of the stars. When adjusted by you to continually track the stars as they move overhead, the alt-azimuth mount produces a zigzag motion, up and down, instead of a smooth arc across the sky.

The equatorial mount also has two perpendicular axes of rotation known as right ascension and declination. Instead of being oriented up and down, it is tilted at the same angle as the Earth's axis of rotation. The equatorial mount comes in two varieties: the *German equatorial* mount, shaped like the letter "T" with the long axis of the "T" is aligned with the Earth's pole, and the fork mount, a two-pronged fork that sits on a wedge that is aligned with the Earth's pole. The base of the fork is one axis of rotation and the prongs are the other.

When properly aligned with the Earth's poles, equatorial mounts can allow the telescope to follow the smooth, arc-like motion of a star across the sky. Also, they can be equipped with setting circles allowing you to easily locate a star by its celestial coordinates, and motorized drives allowing you or your computer to continuously drive the telescope to track a star.

Oculars

Eyepieces, correctly called *oculars*, will, without any doubt, be the first accessory that you purchase after your telescope. Without oculars, you will not be able to see anything of importance or interest. Additional oculars, beyond the one or two supplied with most telescopes, make visual observing much more enjoyable and most variable-star observers have several oculars. Like camera lenses, they determine the field of view and magnification seen through your telescope. Owning a proper selection of oculars adds greatly to the versatility of any telescope. Before buying accessory oculars, I recommend that you take the time to understand some basic facts about these expensive little magnifying lenses that slip into the focuser or star diagonal of your telescope. A little understanding may save you some money too.

Huygens ocular – Designed by Christiaan Huygens (1629–95), this was the first eyepiece made available for use with a telescope. Huygens oculars normally consist of two plano-convex lenses with the flat side pointing at the eye. Today, these type of oculars are mainly used for solar and lunar observations (with the appropriate filter, of course). They have a small apparent field of view (AFOV), usually about 25–40°, and have no color correction. As a rule, these oculars generally have small eye relief.

Huygens–Mittenzwey ocular – A variant of the Huygens eyepiece. The lens that points to the object is replaced by a meniscus. It is a reasonable-to-good eyepiece for slow telescopes (*f*/12 or more). The AFOV is approximately 45–50°. As with Huygens oculars, they generally have small eye relief.

Ramsden ocular – The first achromat, built with two plano-convex lenses, with the convex sides facing each other. They are still found nowadays, especially in small versions (0.965 inches). These oculars suffer from color aberrations but much less than the Huygens and Huygens–Mittenzwey oculars. Ramsden oculars reportedly suffer from internal reflections and usually have a small AFOV with short eye relief.

Kellner ocular – The three-element Kellner is named after Carl Kellner who was born on March 26, 1826 in Hirzenhain, Germany. With a good reputation as a producer of quality optical instruments, Kellner produced oculars for Argelander, among others. His eyepieces were primarily made for use in telescopes but he also produced some for use in microscopes. Eventually, Kellner and his twelve employees built complete telescopes. Until his early death in 1855, his workshop manufactured at least 130 microscopes, five big telescopes and a number of small hand-held telescopes.

Designed in 1849, the eyepiece known today as the Kellner, gives sharp, bright images at low-to-medium powers and is best used on small-to-medium size telescopes. Kellner oculars produce an AFOV close to 40° with good eye relief. As the power of magnification increases, the eye relief will shorten. They are usually considered good, low-cost oculars that are superior to simpler Ramsden and Huygenian designs. As you can see, the Kellner has been around for quite a while and you can find them readily available.

Orthoscopic ocular – The design of orthoscopic eyepieces dates back to the 1800s when Ernst Abbe first designed them to be used for accurate measurements of linear distance on microscope slides. The term *orthoscopic* means an eyepiece that introduces no barrel or pin-cushion distortion. As a result, an object will have the same size when observed anywhere in the field of view. The Abbe design employs a triplet field lens and a singlet eye lens.

Not too many years ago, the four-element *ortho* was considered the best all-around eyepiece available to the amateur astronomer. Today, because of its narrower field of view compared to new designs, the ortho has lost some of this reputation. Orthoscopic oculars produce excellent sharpness, color correction, contrast and have a longer eye relief than Kellners. They are considered great for planetary and lunar observing and are quite good for variable-star observing.

In most cases, the apparent field of view of each of these oculars is approximately 45°. In side-by-side comparisons, fields of view for orthos have been reported to appear equal to or slightly larger than Plössl oculars advertised as having 50° apparent field of view.

Plössl ocular – Considered today's most popular design by many, the four-element Plössl is named after Georg Simon Plössl and provides excellent image quality, good eye relief, and an AFOV of approximately 50°. Plössl was born on September 19, 1794 in Wieden near Vienna and he died on January 30, 1868 in Vienna following a severe injury after dropping a sheet of glass and cutting the artery near his right hand.

High-quality Plössl oculars exhibit high contrast and pin-point sharpness out to the edge and are considered ideal for all observing. *Super-Plössls* have a wider field-of-view than those simply called Plössl lenses. Plössl oculars are considered excellent for variable-star observing.

Erfle ocular – The Erfle lens is named after Heinrich Valentin Erfle who was born on April 11, 1884 in Duerkheim, Germany. His ocular is a five- or six-element optical arrangement and is optimized for a wide apparent field of 60–70°. At low powers, its view has been described as a "picture window." A wide viewing area provides impressive deep-sky views. At high powers, image sharpness reportedly suffers at the edges. Another excellent ocular for observing variable stars.

Wide field ocular – This ocular, with various names, is usually considered the premier ocular today. You will find *super-* and *ultra-wide* eyepieces that incorporate six to eight lens elements to produce apparent fields of view up to 85°. These oculars are best used at low-to-medium power. Wide field oculars have a remarkably wide field of view and you need to move your eye around to take in the whole view. Usually considered the ideal eyepiece for viewing galaxies, star fields and other deep-sky objects, they are superior for observing variable stars since many comparison stars can be viewed within the large field of view. Image quality is excellent, but the number of elements slightly reduces light transmission. This should not really present any concern; however, you *will* pay a premium price for the ultra-wide field of view!

Barlow lens – Not really an ocular but one of the more useful and cost-effective tools for amateur astronomers is a *Barlow lens*. The lens is named after Peter Barlow who was born at Norwich, England in October 1776. The now famous Barlow Lens is the result of a collaboration of Barlow with George Dollond. Barlow calculated a concave achromatic lens that Dollond made in 1833 and mounted to a telescope. Dawes employed it first while measuring close double stars.

A Barlow lens, inserted between the telescope and the ocular, can increase the power of the ocular by two, three and even five times. Of course, when you increase the power of your ocular by two using a Barlow lens, you decrease the field-of-view by two. The trade-off is often fair though. You can double the effective number of oculars that you have by purchasing one Barlow lens. For example, if you own a 40 mm, 25 mm, and 18 mm ocular then a 2× Barlow lens also provides you with a 20 mm, 12.5 mm and a 9 mm lens. Another advantage is eye relief. Looking through a nice big 25 mm ocular is effortless because the eye relief is large. You can keep that same comfortable eye relief and increase the magnifying power of your telescope by simply using a 2 × Barlow. You now have the same eye relief of the larger ocular but you are effectively using a higher-power ocular. Carefully select your oculars so that a Barlow lens doesn't duplicate an ocular already in your possession. In other words, if you have a 32 mm ocular and a 2× Barlow lens, don't buy a 16 mm ocular.

Finder Scopes

Most telescopes come with a finder scope. These are small telescopes attached to your main telescope and they are used to find stars within a larger field of view than your main telescope allows. In most cases, you will find it advantageous to replace the finder scope the came with your telescope with a larger one. A 50 mm, or larger, finder scope will allow you to better match star fields shown on your charts. Large finder scopes are usually more comfortable to use.

Using a Charged Coupled Device (CCD)

CCD stands for charge coupled device. A CCD is a digital camera. The camera contains a chip consisting of a mosaic of light-sensitive electronic micro-cells called photodiodes or *pixels*, a contraction of *picture element*. The chip has a rectangular or square shape and it can be about the size of a dime. The pixel mosaic is called an array. As in photography, you can take exposures lasting several minutes with a CCD since each pixel retains the amount of light received within its memory.

Most astronomical detectors in use today at professional observatories as well as with many amateur telescopes are CCDs. This fact will probably give you the impression that there must be something very special or useful about CCDs. It has been said, almost too much, that CCDs have revolutionized modern astronomy. Without a doubt, they will take their place in astronomical history along with other important discoveries such as the telescope, photographic plates, prisms, and spectroscopy.

For those of you who seek the challenges of CCD observations, you will be pleased to know that they are innumerable. I will not attempt to provide a complete description of CCD operation and CCD use in this book. It would simply be too much for one book.

However, I will provide some things to consider before you make a final decision to move on to CCD observations, followed by a brief description of what a CCD can do when used properly. It seems to me that it is a perfectly natural progression to move from visual observations to instrument observations. Instruments, such as CCDs, allow you to observe things that you will miss when using just your eyes.

First, consider the expense. CCDs can be expensive, at least several hundred dollars and several thousand dollars is normal. Then you probably need additional equipment such as flip-mirrors, filters, and filter mounts. All of this can easily double the cost of just the CCD.

Second, your telescope must be able to track accurately. This means that your telescope must be able to follow a star, and keep it centered on the CCD chip, throughout a long exposure. By long exposure, I mean ten seconds or more. If your telescope cannot do that, it will cost more money to correct that problem.

Third, there is a learning curve – for some it can be quite steep – when you first begin using a CCD. It takes time, patience, more time and much more patience, to begin to properly use a CCD. The challenge is exciting; I am not trying to scare you away. Just be prepared for a time-consuming challenge and budget several weeks to properly learn how to use a CCD for variable-star observing. For some, it may take several months.

The magazine *Sky and Telescope* maintains a Web site specifically for CCD topics. You can find it at ***http:// www.skypub.com/imaging/ccd/ccd.shtml***. At the Web site, you will discover articles such as *Starting out Right in CCD Imaging, Optimizing a CCD Imaging System, Image Processing Basics*, and a few others. This is really a nice place to get a basic start in CCD imaging.

Photoelectric Photometry (PEP)

A stellar photometer is an electrical device that measures the amount of light received from a single star. The process of using a stellar photometer to measure the light intensity of a star is called *photoelectric photometry* and is abbreviated as PEP. People who make photoelectric photometry measurements are sometimes called "Peppers."

There is much interest within the amateur community regarding PEP. For example, the International Amateur–Professional Photoelectric Photometry (IAPPP) group was formed in June 1980 with the goal to facilitate collaborative astronomical research between amateur, student, and professional astronomers. IAPPP provides a medium for the exchange of practical information not normally discussed at symposia or published in other journals. Dr. Douglas Hall, Dyer Observatory, is the publisher and he is well known for his interest in amateur/professional collaboration.

The AAVSO Photoelectric Photometry Program began in 1983 with three observers who in that year contributed 219 observations on 28 stars. Standards were established on how to observe, i.e. taking three measurements of the variable star and one of a check star to monitor constancy of the comparison star. These standards have not changed and the reduction program used today is the same one developed when this program began.

As with using a CCD, there is a learning curve when you begin using PEP methods but it should not be viewed as an obstacle. For more advanced interests, I recommend *Photoelectric Photometry of Variable Stars*, by Douglas Hall and Russell Genet; and *Astronomical Photometry*, by A. Henden and R. Kaitchuck.

Chapter 10

VSO Planning

Make no little plans; they have no magic to stir men's blood and probably themselves will not be realized. Make big plans; aim high in hope and work ...

Daniel H. Burnham

"What variable stars are observable by me at my particular geographical location, during each season, and with my particular equipment?" are also concerns for variable-star observers. A well-developed plan will assist you in answering these questions, and many more, while at the same time directing your activities so that you can begin to prepare to observe variable stars with confidence and little wasted time. Without a plan, you'll struggle while making preparations. And without a plan it's easy to get sidetracked, distracted or waste time looking for stars that are just not in a good position for you to observe. If you believe that you're not the type to become distracted, imagine exploring the sky between Scorpius and Sagittarius, perhaps on a clear night letting your eye wander over near ρ Ophiuchus, or depending upon the season, looping through Orion "the Hunter", across the line of belt stars and on to the Great nebula. If you don't believe that you can't be seduced, at least delayed, when viewing the double cluster in Perseus, the Cygnus loop, the Tarantula nebula in Dorado, or the Great Andromeda galaxy, then you really haven't been behind a telescope's eyepiece in awhile. These are all magnificent sights, typical of what you'll come across while observing, and much like the Siren's song, they can tempt you into altering your course so that you never reach your original destination.

Your plan will help determine your needs for an evening of observation and it will help keep you focused on the specific tasks needed to reach a particular goal. If you need a suggestion as to where to begin, visit the *Variable Star Network*, the *British Astronomical Association Variable Star Section*, or the *American Association of Variable Star Observer's* Web sites[1] to discover what is afoot. See what other variable star observers around the country, or around the world, are observing. These Web sites report the latest variable star interests.

Perhaps a new nova has been detected or a cataclysmic variable has exploded into outburst. Possibly a supernova has been discovered in a bright galaxy or a gamma-ray burst has been reported. Of course, in the end, you should develop your own plan based upon your personal desires. For example, tonight you may wish to time an eclipsing binary, complete 50 visual observations, search for supernovae in 20 galaxies, or make 10 *inner sanctum*[2] estimates. The increased anticipation that you'll feel when preparing your plan can add to the excitement of the actual observations.

An invaluable aid to planning is a log or record book and your log is a great place to record your plan so that you can refer to it during the evening, if necessary. Remember, not everything will go according to plan. As you begin your investigation of the variable stars, you will have successes and you will have failures. A failure is really a success if you look at it as *successfully determining that this isn't the way to do something*, so we'll refer to every happening as a success. It will be your responsibility to determine if any particular success should be repeated or avoided.

Records and logs will allow you to record your successes (those you wish to repeat and those you wish to avoid), observations, unusual occurrences (you're going to be surprised!), memorable events (you're going to be delighted!) and other things too numerous to mention here. Your log can serve not only as a record of your most recent observations but also as a living, historical document that will assist you during the years

---

[1] The Web address for these organizations are provided later in this chapter.

[2] Inner sanctum estimates, a term used by the AAVSO, are those positive visual observations made on stars $13^{m}\!.8$ or fainter, or a fainter-than (you could not see the variable) observation of $14^{m}\!.0$.

to come. It may contain short-cuts and more efficient methods that you discover during observing, notes to yourself regarding problems that you encounter, notes from one season to the next that will allow you to continue observing after a seasonal break of several months, events that you want to remember, ideas that you want to follow-up on later, weather notes, equipment behavior, discoveries, or a multitude of other notable items. I find it enjoyable to review my logs from past years. I'm able to recall great nights, interesting events and I just enjoy reviewing my exploration of the Universe. Not all entries need to have scientific value. Early morning entries, made when I was most tired, are probably best described as entertainment.

Your log or record book is a personal item. There is no preferred format for a log book, so you must develop a format that best suits your personality, observing needs, as well as scientific and historical obligations, if any. The contents of your log can range in style from casual to meticulous; in structure, from direct to elegant; in method, from hand-written to computer generated.

So as to provide a starting point, let's develop a simple log book and we'll begin with an observing plan. An example of a simple observing plan may look something like the following:

VARIABLE STAR LOG - Oct 27, 2001
Nightly observation plan: Checked AAVSO; possible nova detected in Eridanus and V1159 Orionis (SU UMa type CV) is in outburst. VSNET reports a brightening of omega CMa (GCAS type variable). Along with observing these three objects, check the following dwarf novae for outbursts: HL CMa and SU UMa, and estimate the brightness of the following LPVs: R Lep and T Lep. Locate the variable star RU Peg for future observation.

You don't need to make an observing plan every night either. You can make a plan for the month, listing the stars that you want to observe or the galaxies you want to check for supernovae. Then, when you find some unexpected free time, just grab your plan and you're ready to go observing with little time wasted wondering what to view. I've prepared an observing plan for each month, based on the stars visible from my location. Most stars, except those very close to the horizon, are visible for more than a month so you'll have the same stars on several of your monthly plans. The circumpolar stars, those that never dip below the horizon, will be available all year and so will be on every monthly plan. As a suggestion, review your monthly plans each year. Remove those stars that you never observe and periodically add some new ones. Also, consider the Moon and Sun during planning. Observing a variable star near a full Moon or just as the Sun begins to rise can be difficult to do.

Your plan need not be overly complicated but preparing one will help you concentrate on those observations that, for whatever reason, you feel are important and *must* be made during the evening. Once you have accomplished the goals set forth in your plan each evening, you're free to roam around the sky and look at the sights. Don't make your plan too complex or restrictive when you first begin observing variable stars. I've mentioned it before but, in the excitement of the moment, the idea gets lost sometimes; remember, the reason that you're doing this is to observe and have fun. A half-dozen well made observations will leave you with a sense of satisfaction and accomplishment. Add more targets as your abilities improve and as you come to better understand the time commitment. Estimates made in a rush or when you're frustrated and tired may well be less than valueless. Certainly if you are frustrated or unhappy, it won't be fun. It's my opinion, so I could very well be wrong, but I believe that the Universe with all of its attendant stars will be there sufficiently long for you to take your time and enjoy the experience.

Prepare your log or record book ahead of time, perhaps in concert with your nightly observing plan. Label the data columns and sharpen a couple of pencils. I recommend that you use a pencil for your data entry so that an error can be easily corrected. For the first few observing sessions, keep your administrative burden to a minimum. Your primary reason for doing all of this is to observe variable stars. Don't let the paperwork get in your way when you first begin. You will have plenty of time to refine your recording methods.

A simple log contains all of the essential information that you will need to satisfactorily record your variable star observations. I would like to recommend a couple things here. First, whatever format you use for recording the date, stick with it and don't change it from night to night or season to season. Second, record your observations using local time. Trying to convert to Julian Date (JD) or Universal Time (UT) while observing is an additional distraction that is unnecessary. You can convert your observation time to JD or UT later.

A log like this can be kept in a small paper notebook or loose leaf binder, similar to what is used by students for their school work. If you keep your log on a computer hard drive, be sure and consider a back-up copy, such as on a ZIP disk. You'll find that as your logs age, their value increases. Don't risk losing them.

A log containing your observations may look like this:

VARIABLE STAR LOG -- Oct 27, 2001

Nightly observation plan: Checked AAVSO; possible nova detected in Eridanus and V1159 Orionis (SU UMa type CV) is in outburst. VSNET reports a brightening of omega CMa (GCAS type variable). Along with observing these three objects, check the following dwarf novae for outbursts: HL CMa & SU UMa, and estimate the brightness of the following LPVs: R Lep and T Lep. Locate the variable star RU Peg For future observation.

OBSERVATIONS

Date	Time	Star	Mag	Chart	comp star
Oct 27, 2001	10:00PM	RU Peg	13.2	Std RU Peg (d)	126, 135
Oct 27, 2001	10:45PM	V1159 Ori	12.7	Pre V1159 Ori (d)	124, 136
Oct 27, 2001	10:57PM	R Lep	7.7	Std R Lep (b)	75, 78
Oct 27, 2001	11:03PM	T Lep	10.4	Std T Lep (b)	103, 106
Oct 27, 2001	11:09PM	HL CMa	10.9	Pre HL CMa (e)	104, 116 (OUTBURST)
Oct 27, 2001	11:16PM	Omega CMa	3.8	Tycho stars HD 65810 (4.62)	HD 57821 (4.95)

Observer's note: RU Peg is tough to find. Two faint stars are confusing. I really need to check the charts closely. I couldn't find a chart for Omega CMa so I used two stars with Tycho magnitudes. Clouds formed in the sky before I could observe SU Uma.

As you can see, your records should contain information that is considered vital, such as date, time, star observed, estimated magnitude and chart data, but you should also include notes to yourself that will serve as reminders. Personalize your log. Describe in detail, and plain language, what is important during each night of observing. Cryptic notes or unusual abbreviations, made when you're tired, will be difficult to interpret days, weeks or years later.

Notice in this example that we've used local time. When reporting these estimates, you would convert the local time to Julian Date or Universal Time (more on dates and time later in the book). Also, the chart data indicates standard (Std) and preliminary (Pre) charts (more on charts later in the book) and the comparison stars are indicated by just using their magnitudes. You'll notice on the charts that the comparison stars do not have a decimal point. Decimal points, when placed on a star chart, look too much like a star and so are *never* indicated on a chart.

Planning to Report your Estimates

If you desire to report your variable-star estimates, use the proper format for each reporting agency. The AAVSO, VSNET and BAAVSS all have different reporting formats. Check their Web site for the preferred format and use it. Be very careful and report which charts you are using so that the proper comparison stars are recorded. After using several charts, you'll probably notice that a star used as a comparison star may have different magnitude values on different charts. This usually happens when you look at charts from two different organizations such as AAVSO and VSNET. Because this happens, it's important to always use the same chart, use the magnitude shown on the chart, and report the chart used. If everyone uses the same numbers, the data will be consistent.

Selecting Variable-Star Targets

By now, you should feel comfortable with your growing knowledge regarding variable stars and you will probably have little difficulty selecting the variable stars that you wish to begin observing. If I'm wrong and you still feel a bit unsure of yourself, let me help a little. When selecting a variable star that will be suitable for observation, consider these thoughts: *Which stars are visible during the current season? How hard do I want to hunt for a variable star? Are there charts available for the stars that I wish to observe? Is there somewhere that*

I can check my estimates? How often will I be able to get out and observe?

Your first concern should probably be the season. Searching during the month of July for a star that rises in December will be disappointing ... if you're normal. An understanding of which constellations are visible during the various seasons will help you in this regard. A good star atlas is also recommended. Looking for a star positioned at declination –75° while standing in your backyard in Alberta will bring to light the need to understand the celestial coordinate system. Not all stars are visible to all observers standing on the surface of the Earth. That's why observers in Australia, New Zealand, or South Africa observe some stars that are not visible to observers in England, Belgium, Germany or the USA. Of course, depending upon your location, there are many stars that are visible to both southern and northern hemisphere observers during different seasons. For northern observers, Orion is a winter constellation while southern observers know the "Hunter" as a summer constellation.

Along the same line of thought, understanding the limitations of your equipment will allow you to forego the frustration of searching for a 14^m0 cataclysmic variable using a 6-inch telescope. Pursuing faint stars with an instrument of insufficient size is fruitless. Anyway, there is no need to feel you must observe the faintest stars. There are thousands of brighter, poorly studied variables awaiting your scrutiny. Regardless of the size of your observing instrument, you will have access to more stars than you can study in a lifetime.

Think about how hard you want to work at finding a variable too. This may sound a bit confusing but some variables are easy to find while others are more difficult to find. It should be obvious that bright stars are easier to see than fainter ones, but you should also consider stars embedded in dense star fields, variables visible only during outburst, stars visible at the extreme viewing limits of your equipment, or supernovae that look like field stars. These are just a few examples of when variable stars can be difficult to locate. If you want an example, experiment with the challenges of observing the variable stars embedded within the Orion nebula or the open clusters found around the constellations of Scorpius and Sagittarius.

Have you considered variable stars for which no chart exists? With over 36,000 variable stars identified within the *GCVS* alone, and sundry new ones being found each year, you will find many without any chart available. These variables are poorly studied stars and little information, including charts and comparison stars, will be found. Of course it's easier, safer, and quicker to stick with the well-studied variables stars; however, much enjoyment can be found if you leave the well traveled road. I think Robert Frost said it best,

Two roads diverged in a wood, and I –
I took the one less traveled by,
And that has made all the difference.

After making your first few estimates as to the brightness of a variable star, you will begin to question your accuracy. It's human nature to doubt and, anyway, you *should* be concerned with errors and make every effort to ensure that your estimate is as accurate as possible. No one wants to record errors, and reporting gross inaccuracies is embarrassing. Doubt is the Universe's way of reminding you to check your work. The *Variable Star Network* (VSNET), provides nightly observation reports for hundreds of variable stars showing what other observers around the world are estimating. When you first begin, you should be practicing on well-known stars and by doing so you will have no trouble finding the latest observation reports for hundreds, perhaps thousands, of well-known variable stars with which you can compare your estimates. As you develop experience, you'll find less need to check and see what others are reporting. Eventually, your confidence will be such that, in your opinion, only your estimate is correct; everyone else is a little bit off.

Begin to understand the time commitment required for observing variable stars. When you first start observing variable stars it will be effortless to make time to get out under the night sky and observe. You'll have superhuman strength, you will be indefatigable, dawn to dusk observing will seem effortless, cold air will be refreshing, going to work after a 90-minute nap will be an acceptable trade. A couple of weeks later, reality intrudes. Develop a sustainable effort, one that keeps you interested but doesn't lead to an impossible undertaking. Remember, the stars will be there for years to come. Perhaps longer.

With these thoughts in mind, you are ready to begin to make some serious plans. It's time to prepare a list of targets and march boldly into the darkness, seize the night and observe variable stars!

Planning Resources

Here are a handful of basic resources that you can use to assist you in your planning. Many others, some quite comprehensive, exist and will become useful as your experience grows. If all of this seems like too much trouble right now and you just want to get out and look at some variable stars, forget all of this and go outside. Eventually you'll discover the need for a little planning but don't let it get in the way of some fun right now. You'll be back to work on your plan.

The *Combined General Catalog of Variable Stars (GCVS)* – This is considered an essential document to variable-star observers. You don't need a printed copy in your bookshelf since it's available on the Internet (***http://lnfm1.sai.msu.su/GCVS/***). The *Combined General Catalog of Variable Stars* is the primary source of information on variable stars. The current *GCVS* contains the combined computer-readable version of the *GCVS*, Vols. I–III (Kholopov *et al.* 1985–8) and *Name-Lists of Variable Stars Nos. 67–75*. The total number of designated variable stars has now reached 36,064. Within this catalog you will find the variable stars listed by their names, position in the sky, brightness (magnitude) at maximum and minimum, spectral and luminosity types (when known), and much more information. When beginning to observe variable stars, this is *the* catalog to use.

Bright Star Catalog (***http://amase.gsfc.nasa.gov/amase/MissionPages/YALEBSC.html***) –This catalog is not used much as a source for variable-star data, but it is widely used for basic astronomical and astrophysical data for stars brighter than $6^{m}.5$. Within the catalog you will find 9110 objects of which 9096 are stars. There are many stars listed as variable found within this catalog.

The *Variable Star Network* (VSNET) – The VSNET is a comprehensive Web site maintained by professional astronomers from Kyoto University, Japan (***http://www.kusastro.kyoto-u.ac.jp/vsnet/index.html***). The Web site is in English and you will find information on many variable stars here, along with the latest stars of interest.

TA/BAAVSS Recurrent Objects Programme – The Astronomer and the British Astronomical Association Variable Star Section (***http://www.britastro.org/vss/***) maintains a recurrent object program consisting of variable stars that have been well observed. You'll also find an eclipsing dwarf nova program, eclipsing binary program, and binocular observation program.

Information Bulletins on Variable Stars (IBVS) – The *Information Bulletins on Variable Stars* (**http://www. konkoly.hu/IBVS/IBVS.html**) are published by Konkoly Observatory , Budapest, Hungary. These bulletins are an excellent resource for amateur astronomers and you will find much information regarding many variable stars. Another Web site that you must visit.

AAVSO Bulletin – The American Association of Variable Star Observers Bulletin (**http://www.aavso.org/ bulletin/**) contains predicted dates of maxima and minima of long-period variables in a schematic representation and shows when a variable will be brighter than magnitude 11.0 or fainter than magnitude 13.5. Along with the bulletin, you will find several other publications that will assist you in planning a night of observing. The following publications are available on-line and can be downloaded from the AAVSO site:

AAVSO Manual for Visual Observing of Variable Stars – This is a good guide to variable-star observing. It incorporates a lot of the basic information from the *Manual for Observing Variable Stars*, that was published in 1970, as well as information from various AAVSO observing publications that have developed since then.

Catalog of variable star charts – Several types of charts are available including *constellation finder charts* presented in wide-field plots that encompass an entire constellation, *standard charts* for variable stars that have been in the AAVSO visual observing program for decades, *preliminary charts* for variable stars that have comparison star sequences that many not be well established and *special-purpose charts* such as those used for observing eclipsing binary or RR Lyrae stars or for observers with photoelectric photometers or CCD cameras.

AAVSO Alert Notices are published irregularly and these serve to alert those interested to the discovery of novae, unusual activity of variable stars and requests from astronomers for simultaneous observations.

The *Eclipsing Binary Ephemeris* is available and shows the predicted time of mid-minimum for eclipsing binaries in the AAVSO Eclipsing Binary observation program.

The *RR Lyrae Ephemeris* is available and shows the predicted time of maximum for RR Lyrae variable stars in the AAVSO RR Lyrae Stars observing program.

Supernova Search Manual, 1993, written by Robert O. Evens, Coonabarabran, NSW, Australia is available and is considered an excellent manual for supernova patrols.

The Catalog and Atlas of Cataclysmic Variables, Living Edition – This is a marvelous resource for cataclysmic variable observers (***http://icarus.stsci.edu/~downes/cvcat/***); a Web-based version of the previously published catalogs (Downes and Shara 1993, *PASP* **105**, 127; Downes, Webbink, and Shara 1997, *PASP* **109**, 345). Over a thousand CVs are listed at the Web site.

Your log or record book – As it matures, your log or record book will become a great resource that you can use during your planning. Look for any unfinished projects, notes that indicate your interest in something for which you didn't have the time to explore or objects seen but not indicated on a chart. These are reasons it's a good idea to write down everything.

Now it's time to use your plan to make the necessary preparations so that you can observe variable stars.

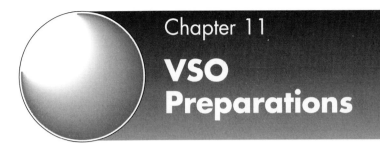

Chapter 11
VSO Preparations

Somebody said that it couldn't be done,
But he with a chuckle replied
That "maybe it couldn't," but he would be one
Who wouldn't say so till he'd tried.
So he buckled right in with the trace of a grin
On his face. If he worried he hid it.
He started to sing as he tackled the thing
That couldn't be done, and he did it.

Edgar A. Guest

"Are preparations different from planning?"

A *plan* is a detailed method or arrangement of details for the accomplishment of a project. *Preparations* are the measures that serve to make ready for a specific purpose. Without a good plan, your preparations will probably be found to be incomplete since you're not going to have a well-defined goal; you're not going to know exactly *where* you're going, *what* you need to get there or how to tell *when* you get there. Without proper preparations, you're not going to have available the equipment or special items that you will need to execute your plan.

It really isn't that difficult or time-consuming. A little research, in the form of a plan to lay the groundwork for your preparations and a solid understanding of your goals, are the bare essentials. You don't need an advanced degree in astronomy or astrophysics to enjoy observing variable stars any more than you need an advanced degree in engineering to enjoy driving a well-built automobile or a degree in sports medicine to enjoy physical fitness. And you don't need to spend a good portion of your yearly income to have the necessary equipment. To observe variable stars all that is required are desire, binoculars or a small telescope, a good plan and adequate preparation.

I want to make some suggestions regarding preparation, and in doing so I hope to increase the time you can spend actually observing something that is interesting. Preparation seems to be a serious stumbling block, after deciding what equipment to purchase, when it comes to really getting out under the stars and doing something. Speculating as to what you'll need each evening can become tedious; perhaps you're wondering if you should take every piece of astronomical equipment that you own or leave some stuff in the house or in the car (what items though?), how should you dress, do you need food or something to drink, how about extra batteries, how will you carry everything, will things get lost in the dark, do you need electrical power? You can be paralyzed into doing nothing if you don't approach your efforts with some method in mind.

If you're not sure where to begin, refer back to your plan for guidance. During your planning process, you've considered which stars are available to you, based on the season, your geographical location and the capabilities of your equipment. By doing so, you've identified a good number of variable stars that are possible to observe. You should have a list of stars available for your viewing. Its time to think about what equipment you will need to meet the objectives set forth in your plan. Simply apply some common sense.

Think about your equipment first. Setting up a one hundred pound (50 kg) Dobsonian telescope to observe β Persei doesn't make much sense because it's a bright star, best viewed with binoculars. On the other hand, preparing to observe BY CMa, a 13^m0 Cepheid variable, using 10×50 binoculars will be disappointing. Take only the equipment needed to observe the stars listed in your plan. You may not need all of your oculars; take only those needed. Do you need batteries or an electrical outlet to run your telescope? Do you have a flashlight? Red filter?

Consider your dress carefully. If it's January and you live in Belgium, you had better think of your feet, hands, and head. Good boots, thick socks, gloves or mittens and a hat are essential; they are not luxury items. On the other hand, during August in New Mexico, it's still 80°F (25 °C) at midnight. Shorts and a light shirt will make the evening more enjoyable.

Food and something to drink can make a long evening of viewing more enjoyable. A sandwich or two, some cookies, or some fruit will provide the energy needed for several hours of observing and an empty stomach won't distract your viewing. Of course, this can cause problems too. Melting chocolate bars, especially when using someone else's equipment, should be avoided. Alcoholic drinks are not a good idea for a number of reasons but water is always a good idea. Greasy chips, fried chicken and corn-on-the-cob are probably best left at home. I'm sure that you get the idea.

Storage and carrying cases are well worth the cost. Not only do they make carrying everything a little easier, it's easy to develop a loading plan, designating a specific place for each item, that will also allow you to check and ensure nothing is left behind at the end of the evening. Realizing that you've left your best Nagler ocular on the desert floor is an excellent way of inducing an adrenaline high. Make a checklist and list everything that you take for an evening of observations. When it's time to come home, get your checklist out and check everything. That's why it's called a checklist.

Initially, you'll want everything that you own with you when you observe. As you spend more time observing, you'll revise your list of items and begin to select those things that you really need. It gets easier as time passes.

The Celestial Coordinate System

One of the things that you will absolutely need is a star chart or atlas. Before we actually look at a star chart or atlas, there is one thing that you must understand: the celestial coordinate system. This is an imaginary projection of the Earth's geographical coordinate system onto the celestial sphere that seems to turn overhead at night. This celestial grid is complete with equator, latitudes, longitudes and poles.

As you know, the Earth is in constant motion as it rotates on its axis. As a result, the celestial coordinate system is being displaced very slowly with respect to the stars. The celestial equator is a 360-degree circle bisecting the celestial sphere into the northern celestial hemisphere and the southern celestial hemisphere. Like the Earth's equator, it is the prime parallel of latitude and is designated 0 degrees.

The celestial parallels of latitude are called *coordinates of declination* (Dec) and much like the Earth's latitudes they are named for their angular distances from the celestial equator, measured in degrees, minutes and seconds of arc. There are 60 minutes of arc in each degree, and 60 seconds of arc in each arc minute. Declinations north of the celestial equator are designated "+" and declinations south are designated "–." The celestial North Pole is located at +90 degrees and the celestial South Pole is located at –90°.

The celestial meridians of longitude are called *coordinates of right ascension* (RA) and like the Earth's longitude meridians they extend from pole to pole. There are 24 major RA coordinates, evenly spaced around the 360° equator, one every 15 degrees. Like the Earth's longitudes, RA coordinates are a measure of time as well as angular distance. We speak of the Earth's major longitude meridians as being separated by one hour of time because the Earth rotates once every 24 hours (one hour = 15 degrees). The same principle applies to celestial longitudes since the celestial sphere appears to rotate once every 24 hours. Right ascension hours are also divided into minutes of arc and seconds of arc, with each hour having 60 minutes of arc and each arc minute being divided into 60 arc seconds.

Astronomers prefer the time designation for RA coordinates even though the coordinates denote locations on the celestial sphere because this makes it easier to tell how long it will be before a particular star will cross a particular north–south line in the sky. So, RA coordinates are marked off in units of time eastward, from an arbitrary point on the celestial equator in the constellation Pisces. The prime RA coordinate which passes through this point is designated "0 hours 0 minutes 0 seconds." We call this reference point the vernal equinox where it crosses the celestial equator. All other coordinates are names for the number of hours, minutes and seconds that they lag behind this coordinate after it passes overhead moving westward.

Given the celestial coordinate system, it now becomes possible to find celestial objects by translating their celestial coordinates using telescope-pointing positions. For this you use setting circles for RA and Dec to find celestial coordinates for stellar objects which are given in star charts and reference books.

Charts

One of the most daunting tasks new observers encounter is actually locating variable stars in the sky. Although finding the region of the sky where the variable resides may be straightforward, the actual identification of the variable is a learned skill involving patience, persistence, and a good chart. You will find that several types of charts exist.

First, you'll need *finding charts*. The purpose of a finding chart is to get you to the approximate vicinity of the variable star that you want to observe. A good star atlas can serve as a finding chart. Learn the bright stars, visible to your naked eye.

To aid observers with the identification of the variable star, many organizations such as AAVSO, BAA VSS, VSNET and ASSA provide charts. Figure 11.1

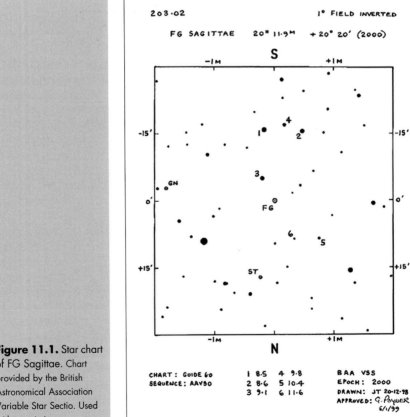

Figure 11.1. Star chart of FG Sagittae. Chart provided by the British Astronomical Association Variable Star Sectio. Used with permission.

is a chart for FG Sge provided by the BAA VSS. You will notice that north and south are inverted and that several stars are indicated by a number. At the bottom of the chart you will see the magnitude listed for the comparison stars. Visit several Web sites from the different organizations. Take a look at what they offer. Right now, I'm going to use the AAVSO as an example.

Like many organizations, AAVSO provides finding charts at their Web site. These charts display the field of the variable, along with other pertinent information that may be useful when observing. Shown on the charts are stars of known constant magnitude, referred to as comparison stars, that are used to make brightness estimates of the variable. With over 3000 charts, the AAVSO is one of the major sources of charts of variable stars. All of these charts are currently available online and may be downloaded for free, or may also be purchased for a fee through AAVSO.

The AAVSO provides different types of charts tailored to meet the needs, experience, and programs of their observers. When making variable star estimates for the AAVSO, they require observers to use these charts in order to avoid the conflict that can arise when magnitudes for the same comparison star are derived from different sets of charts. This could result in two different degrees of variation being recorded for the same star.

Constellation finder charts present wide-field plots that encompass an entire constellation with the location of bright stars and selected variables charted (Figure 11.2). Originally produced for their Hands-on-Astrophysics educational project, these charts may also be of use to the beginner trying to find their way around the sky.

Standard charts are for variable stars that have been in the AAVSO visual observing program for decades, and have comparison star sequences that are established and not subject to change. Always use a standard chart whenever possible. Any new observer should begin by using standard charts.

Preliminary charts are for variable stars that have comparison star sequences that may not be well established, and thus, are subject to change (Figure 11.3). These charts are typically for the more experienced observer.

Also available are *reversed charts*. These charts are provided in both standard and preliminary format and have been reversed north to south for use with

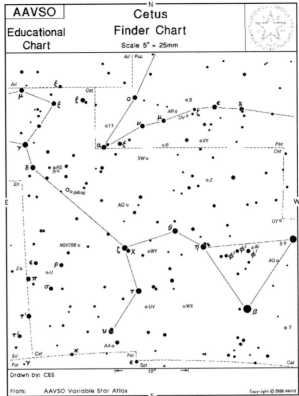

Figure 11.2. Finder chart for the constellation Cetus. Chart provided by the AAVSO. Used with permission.

telescopes with an odd number of reflections, such as Schmidt–Cassegrains or refractors with diagonal mirrors. At present, reversed charts do not exist for all variables charted in the AAVSO visual observing program. A chart reversal project is underway.

Special-purpose charts such as those used for observing eclipsing binaries or RR Lyrae stars or for observing with photoelectric photometers or CCD cameras are also available. You'll also need supernovae charts if you're going to be hunting these interesting events. Supernovae charts are really charts of galaxies with the field stars indicated so that you will recognize a supernova when it appears (Figure 11.4). It will be the star not shown on the chart. If you detect a supernova, you must determine its location with a good amount of precision. Usually, the chart will not be a good source of accurate position. Also, remember to get a confirmation from someone else. The Internet is a good place to announce your discovery and ask for a confirmation observation from another astronomer.

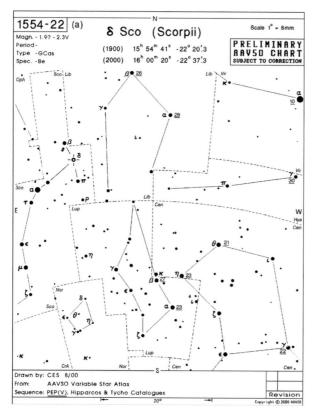

Figure 11.3.
Preliminary chart for δ Sco. Chart provided by the AAVSO. Used with permission.

The charts range in scale from 5 arc minutes per millimeter, "a" scale charts, to 2.5 arc seconds, "g" scale charts. The scales needed for your observing program will depend on the equipment that you are using. Table 11.1 summarizes this information.

Again, many organizations provide charts. It will be well worth your time and effort to take a look at what they offer. You will find the Web address for several organizations later in the book.

Preparing Your Own Charts

Some of the astronomical associations are not going to be happy with this section of the book (they really like you to use their charts) but for you to truly understand how a star chart works you must make and use a few of your own. Besides, in some cases, you'll find no charts

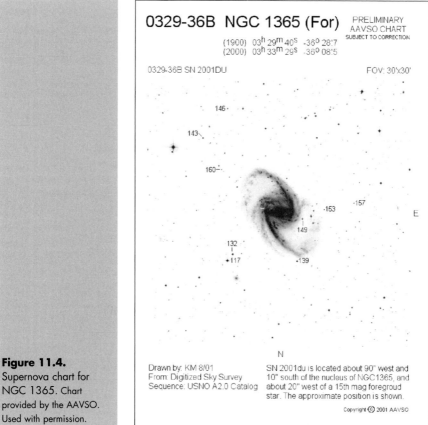

0329-36B NGC 1365 (For) PRELIMINARY AAVSO CHART
SUBJECT TO CORRECTION

(1900) 03^h 29^m 40^s -36° 28'.7
(2000) 03^h 33^m 29^s -36° 08'.5

0329-36B SN 2001DU   FOV: 30'X30'

146
143
160
                                    -157
                        -153
                                                E
                        149
        132
       117   -139

                        N

Drawn by: KM 8/01   SN 2001du is located about 90" west and
From: Digitized Sky Survey   10" south of the nucleus of NGC1365, and
Sequence: USNO A2.0 Catalog   about 20" west of a 15th mag foregroud
                               star. The approximate position is shown.

Copyright © 2001 AAVSO

Figure 11.4.
Supernova chart for
NGC 1365. Chart
provided by the AAVSO.
Used with permission.

available for a particular star or field and so you must prepare your own.

A great place to begin to produce your own finder charts is at the USNO Finder Chart Service (***http://ftp. nofs.navy.mil/data/FchPix/***). This facility allows you to extract catalog data from the USNO-A2.0 and/or ACT

Table 11.1.

Scale	Arc/min	Area	Recommended use
a	5 min	15 degrees	binoculars/finder
ab	2.5 min	7.5 degrees	binoculars/finder
b	1 min	8 degrees	small telescope
c	40 sec	2 degree	3–4 inch telescope
d	20 sec	1 degree	> 4 inch telescope
e	10 sec	30 minutes	large telescope
f	5 sec	15 minutes	large telescope
g	2.5 sec	7.5 minutes	large telescope

catalogs and plot finder charts from these lists. In addition, you can find the raw images from scanning the major photographic surveys. You can overplot the catalog data onto the images, as well as overplot your own additional markers.

When you first go to this Web site, you will be provided with a split screen. The top part will provide instructions for using the Finder Chart Service. The lower screen allows you to query the service. This is a remarkable service and the folks at the US Naval Observatory provide a great resource; however, because the Finder Chart Service is a comprehensive service, you're not going to simply push a button and get a nice finder chart. Plan on spending an hour or so investigating this facility. Don't go to this service minutes before you need a finder chart and expect to zip through the process and get your chart.

Another great place to begin to put a chart together is SIMBAD at the *Centre de Données Astronomiques de Strasbourg* (CDS) (*http://simbad.u-strasbg.fr/sim-fid.pl*). The SIMBAD astronomical database provides basic data, cross-identifications and a bibliography for astronomical objects outside the solar system. The "SIMBAD: Query by identifier, coordinates or reference code" page allows you to enter a variable star by name or coordinates. It then provides you with a data page. From the basic data page, you can open the "ALADIN Java: Sky Atlas." ALADIN is an interactive software sky atlas that allows you to visualize digitized images of any part of the sky, to superimpose entries from astronomical catalogs, and to interactively access related data and information.

Sound complicated? The CDS provides a tutorial for their main services, including the SIMBAD astronomical database and ALADIN sky atlas, at (*http://cdsweb.u-strasbg.fr/Tutorial/index.gml*).

Star Sequences

A star sequence is a list of stars used to judge the brightness of a variable star. They are the comparison stars and they must be selected very carefully. There are several things to consider when selecting comparison stars.

First, they cannot be variable. Finding stars that are not variable is not as simple as it sounds. In fact, it's not unheard of to discover that a comparison star that has been used for some time is actually variable when it's examined closely. Also, when observing the brightest stars, it's fairly difficult to find close bright stars that are not variable.

Then, you want a number of constant stars that span the amplitude range of the variable. You want a sequence of stars that extend a little bit brighter and fainter than the variable is known to shine. This way, you can interpolate by placing your variable *between* a brighter and fainter star when estimating its brightness.

In an ideal sequence, all of the stars are of the same spectral type as the variable. Unfortunately, this will rarely happen. If possible, select stars close to the same spectral type as the variable. Even this may be difficult to do. For example, if the variable star is an A-type star, try to find other A-type stars to use as comparison stars. When this is impossible, try to use B-type or F-type stars. Stay as close to the variable star's spectral type as possible.

And finally, you want your comparison stars to be close enough to the variable so that you can make quick comparisons without moving long distances across the sky. Not only is this time-consuming but if you are comparing your variable with a comparison star more than a degree removed from the variable, the effects of the atmosphere begin to become noticeable.

As you can see, much effort goes into selecting the star sequence that provides the comparison stars. The importance of comparing your variable star to acceptable comparison stars should also be obvious now. If hundreds, even thousands, of variable star observers all compared their variable stars to different comparison stars, with no concern for the factors just discussed, it would be impossible to compare estimates, check for errors, or standardize the observations.

When possible, use approved charts from the BAA VSS, AAVSO or VSNET. Other organizations provide charts too. When you're required to make your own charts, check the literature for stars that have been used as comparison stars in the past. If you find yourself needing comparison stars and none exist, carefully develop your own sequence with the assistance of other variable-star observers. The odds are, for even the smallest portion of the sky, someone, somewhere has a good start on a star sequence that will provide comparison stars.

Dates and Times

Astronomers use a different time system than mere mortals. Amateur astronomers need to use these time systems too.

As I've said before, use the local time when making entries in your log. Converting your local date and time to one of the astronomical systems should be done *after* you're finished observing. Your full attention should be focused on accurate observations.

Universal time (UT) is used by all astronomers. UT is simply the time in Greenwich, England. Instead of a 12-hour clock, a 24-hour clock is used. UT allows astronomers from around the world to have a common reference for time. If we all know that something *astronomical* is going to happen at 22:43 hrs UT, we simply need to convert it to local time. Where I live, we are six or seven hours behind UT, depending upon whether daylight savings time is in effect, so I simply subtract the correct number of hours from UT to determine local time for me. When I convert local time to UT, I just add the correct number of hours. After you've used this system for awhile, it's second nature.

The Julian date (JD) is used by variable-star observers and was designed by Joseph Justus Scaliger in 1582 and has nothing to do with the Julian calendar. Scaliger named his system in honor of his father, Julius Caesar Scaliger.

This system enables variable-star observers to more easily compare the characteristics of stars over periods of years. The first Julian day begins at noon in Greenwich on January 1, 4713 BC. This date was selected because it happened to mark the start of three independent cycles of solar and lunar phenomena.

You're probably wondering if you need to figure the number of days since 4713 BC. No, you don't. A list of Julian days, essentially a calendar, is available from the major variable star organizations. For your convenience, noon on 1 January 2002, is Julian day 2452276.0 It's easy to set up a list of Julian days using a spreadsheet. Remember, the Julian day begins at noon in Greenwich; in other words, 1200 hrs UT.

The Julian day also allows you to indicate the time of day when you record or report an observation. Notice the decimal position for the Julian day in the last paragraph. Usually, the type of variable star being observed dictates the accuracy, that's the number of decimal places that you need to record. For example, if you are observing an LPV, such as R Lep (Hind's Crimson star) with a period of 427^d, then simply recording the day of your observation is sufficiently accurate. One day out of 427 is an accuracy of 0.2%. On the other hand, nearby S Eri, an RRc variable, with a period of $0^d.273$ (6 hrs, 33 min, 6 sec) needs to be observed much more frequently and each observation must be reported with a greater precision than is required for R Lep. Since S Eri's period is accurate to a thousandth of a day (3 decimal places) you must report your observations with at least the same accuracy.

A calculator will help or you can set up a spreadsheet to help you. I find it easy to work with seconds when it comes to short intervals of time. If you multiply the number of hours in a day by 60 you can calculate the number of minutes in a day. The answer is 1440. If you now multiply the number of minutes in a day by 60 you can calculate the number of seconds in a day. The answer is 86,400. Now you can see that S Eri's period is 0.273 times 86,400, or 23,587.2 seconds. You can easily convert this to hours and minutes.

When it comes time to calculate the decimal time portion of the JD, perhaps the moment of your observation, follow a few simple procedures. For example, if your observation is at 11:34 PM, first convert the time to UT. Let's say that UT is six hours ahead of you. Simply add six hours to your local time. The time of your observation, 11:34 PM local time, is 0534 hrs UT.

Remember, the JD starts, and ends, at noon UT. Let's say that your observation was made on January 2, 2002. The JD would then be 2452277.0. Now refer back to the UT time of your observation. Since the JD begins and ends at noon, 12 hours must be added to the UT. So, 0534 plus 1200 equals 1734 hrs JD. Convert this to seconds by multiplying the 17 hours by 60 and then by 60 again. The answer is 61,200. Then multiply the 34 minutes by 60. The answer is 2040. Add 61,200 and 2040 for the total seconds. The answer is 63,240. Divide this by 86,400 to get the decimal time. The answer is 0.73194. Round your answer to an appropriate number.

Any questions on why you should be doing this after you have finished observing?

The Internet

The Internet provides you with a doorway to the virtual Universe. As an aid to your preparations, the following Web sites are recommended. One of the really neat things about the Internet is that you can still view the sky on cloudy, windy, rainy days. Check these Web sites. You're going to be surprised.

*The STScI Digitized Sky Survey (**http://archive.stsci. edu/dss/**)* The Digitized Sky Survey comprises a set of all-sky photographs conducted with the Palomar and UK Schmidt telescopes. The Catalogs and Surveys Branch (CASB) is digitizing the photographic plates to support HST observing programs but also as a service to the astronomical community, including amateur astronomers. Images of any part of the sky may be extracted from the DSS, in either FITS or GIF format.

*Centre de Données Astronomiques de Strasbourg (CDS) (**http://cdsweb.u-strasbg.fr/**)* The Strasbourg Astronomical Data Center (CDS) is dedicated to the collection and worldwide distribution of astronomical data and related information. It is located at the Strasbourg Astronomical Observatory, France and hosts the SIMBAD astronomical database, the world reference database for the identification of astronomical objects.

The CDS' goals are to: collect all of the useful information concerning astronomical objects that is available in computerized form – observational data produced by observatories around the world, on the ground or in space; upgrade these data by critical evaluations and comparisons; distribute the results to the astronomical community; and conduct research, using these data.

*United States Naval Observatory, Flagstaff Station (USNO) (**http://www.nofs.navy.mil/**)* One of the most generous and dedicated professional astronomers, when it comes to sharing his time with amateurs, works at the United States Naval Observatory, Flagstaff Station (USNOFS). Dr. Arne Henden has been assisting amateur astronomers for many years.

This facility allows you to extract catalog data from the USNO-A2.0 and/or ACT catalogs and plot up finder charts from these lists. In addition, they have available the raw images from scanning the major photographic surveys. You can overplot the catalog data onto the images, as well as overplot your own additional markers.

USNO-A2.0 contains entries for over a half billion stars (526,230,881, to be exact!) that were detected in the digitized images of three photographic sky surveys. USNO-SA2.0 is a subset of USNO-A2.0 which is a lot easier to handle on a small computer because it contains only a tenth as many stars as the parent catalog (54,787,624 stars in all).

HIPPARCOS/TYCHO (***http://astro.estec.esa.nl/Hipparcos/***) This site could, with little argument, be considered a gold mine of recoverable variable-star data. Hipparcos is a pioneering space experiment dedicated to the precise measurement of the positions, parallaxes and proper motions of the stars. The intended goal was to measure the five astrometric parameters of some 120,000 primary program stars to a precision of some 2 to 4 milliarcsec, over a planned mission lifetime of 2.5 years, and the astrometric and two-color photometric properties of some 400,000 additional stars (the Tycho experiment) to a somewhat lower astrometric precision.

The Hipparcos and Tycho Catalogs contain a wealth of information in a user-friendly format. Available in both printed and machine-readable form, these catalogs can be exploited by both professional and amateur astronomers.

Astrophysics Data System (ADS) (***http://adswww.harvard.edu/ads_articles.html***) The Astrophysics Data System (ADS) can be thought of as your own private professional publication library. This service provides free and unrestricted access to scanned images of journals, conference proceedings and books in astronomy and astrophysics.

Los Alamos National Laboratory (LANL) (***http://xxx.lanl.gov/***) The Los Alamos National Library (LANL) pre-print service will allow you to read professional research articles before they are actually released in the various journals such as *The Astrophysics Journal, Publication of the Astronomical Society of the Pacific, The Astronomical Journal*, and many others. This is a great service and should not be ignored but be warned, these are professional research articles and the content of the articles are written at the professional research level.

Chapter 12

VSO Methods

Things won are done, Joy's soul lies in the doing.
William Shakespeare

Grab your binoculars or telescope, your log or record book, as well as your star charts or star atlas and move outside, under the stars. As Mr. Shakespeare says, "Joy's soul lies in the doing."

Find a comfortable location that provides you visual access to the sky. Try and find a place that protects your eyes from glaring light. Street lights, security lights from your neighbor, and house lights coming out through windows are usually the types of light that will cause you some frustration. If you can move yourself out into a remote location, far from city lights, that will be great. If you can't, use a tree or building to block the light. It doesn't have to be perfectly dark for you to enjoy yourself. There can even be some clouds in the sky.

Get comfortable. Use a blanket on the ground, a soft patio chair or a fancy observing seat purchased specifically for astronomical observing. Use a pillow if necessary. The important thing is to get you head and eyes oriented to the sky so that you are not straining or uncomfortable. I'll mention this again.

After referring to an atlas or sky chart, look up and locate the field or region of the sky in which the variable star of interest is located. Again, as during the planning process, knowing the constellations will be very helpful. Take out your wide-field chart and orient it so that it matches what you see in the sky. Concentrate on the brightest stars first. Remember, when looking at a chart placed on your lap, east and west are reversed relative to the way you normally distinguish them. Also, on some charts, north and south are reversed. When you first start, hold your chart over your head, against the sky. It make orientation easier.

Your first surprise will be that you'll notice the stars in the sky look different from what you expect when compared with your atlas or chart. My guess is that when you first begin using a chart and your eyes leave the chart and slowly rise toward the sky ... you will become lost! Don't become alarmed. It's all just a matter of perspective. It won't be long before you make the mental adjustments automatically and it will seem normal. Give it time.

Once you get out under the sky, spend fifteen minutes or so just looking around. Let your eyes wander through the Milky Way near Scorpius and Sagittarius, Centarus, Orion or Leo, depending upon the season and latitude. Taking a few minutes to look around will allow your eyes to adapt to the night sky too. In just a short time, you will see many more stars than when you first walked outside. The point is, don't be in a hurry. Take the time to learn the positions of the constellations and stars. And don't forget to enjoy the sky; this is after all a hobby and you're doing all of this for enjoyment. Professional astronomers don't enjoy the freedom that you have as an amateur. Don't take it for granted.

Variable Star Observing with Binoculars

Regardless of what equipment you have, many variable-star observers recommend that you begin observing variable stars with binoculars. This sounds like pretty good advice. If you have a telescope you won't want to do it though. I understand, but here is why most observers recommend starting with binoculars.

First, it's easy to set up quickly and start observing with the least amount of fuss when using binoculars. You're going to get positive feedback quickly with very little effort. All that you will be required to do is lay back, keep your eyes open, breath and look.

Secondly, when the time comes, you'll have many more comparison stars using binoculars. The field of view for binoculars is greater than a telescope's so naturally you'll be looking at a bigger portion of the sky. More sky means more stars.

Setting up for binocular viewing is simple and straightforward. You will need binoculars, star charts/atlas, a record book, a pencil and a red filtered flashlight. It's important to be comfortable when observing variable stars. Your attention needs to be on the stars, not on a sore neck or back. Lay on a blanket covering the ground or sit in a yard chair of some type. Use a pillow. Again, the critical thing here is that you must orient your head so that it is naturally tilted up toward the sky. If you don't do this, you will fatigue quickly; worse case, you're going to get a cramp or a headache. Most people will not pursue a hobby in the hopes of developing a cramp or a headache.

Scan the sky with your binoculars. Find the bright stars. Notice the colors. If you're a normal human being, this should be a thoroughly amazing experience. Think about what you are seeing. This is the Universe that you are viewing!

If looking up doesn't suit your fancy, look down. There are devices available in which you look down into a large mirror. The mirror reflects the sky from above. Looking down into a large mirror with binoculars will give you a nice view of the sky above without straining your neck. The disadvantages are that the mirror will not reflect 100% of the light from the stars so you'll lose a little brightness, and the mirror, no matter how it's positioned, will be susceptible to wobble. You're already using binoculars that, by their very nature, tremble so any additional quiver can be annoying. Attach a low-priced laser pointer to your binoculars, using rubber bands, and you can use it to direct your point of view. A little adjustment will be necessary in the beginning to center the laser with your field of view.

You can spend months or years observing with binoculars; maybe even a lifetime. As time goes by, you'll gain much experience observing variable stars with your binoculars. You'll learn the constellations and asterisms, bright stars, nebulae and star clusters that will allow you to move around the sky with confidence. The knowledge that you develop will come in handy in the years to come. And don't forget to enjoy the journey, there is no final destination.

Variable Star Observing with a Telescope

Even when a telescope is your primary instrument, you should have binoculars nearby and use them. It's easier to scan the sky and search for a particular star field with binoculars, then switch to your telescope when you've found something of interest. The binoculars will give you a wider field of view so it's easier to figure out where you're looking.

Again, comfort is important. Assume a comfortable viewing position with your head and eyes level with your eyepiece. Make sure that you can reach your telescope controls so that you can make minor adjustments. Keep your charts and log book within arm's reach.

The big difference with using a telescope compared to using binoculars is that you will be observing using only one eye. This can be challenging since, normally, you spend most of your time using both eyes to see. If you find it difficult to keep one eye closed while observing, using an eye patch will help. Place the eye patch over your non-observing eye and you won't need to be concerned with what that eye is doing. It's usually a little more comfortable keeping both eyes open, and using the patch to cover your non-observing eye allows you to do so.

Depending upon the type of mount that is supporting your telescope, some positions will be more difficult to reach than others. For example, when using a German Equatorial mount, especially when supporting a re-fractor telescope, observing directly overhead can be difficult because your eyepiece will be located closer to the ground than in any other orientation. If you have a fork mount and you are not using a *wedge*, observing directly overhead is also difficult. A wedge allows an alti-azimuth mount to be converted into an equatorial mount. Dramatic changes in position may require you to rebalance your telescope and it's not unusual for this to occur several times during an evening. Dobsonian telescopes do not produce these annoyances. They're easy to move to new positions and the eyepiece is usually in a good position.

All of these preparations, whether it be with binoculars or a telescope, are made in an effort to eventually do one thing – accurately estimate the magnitude of a variable star. Let's examine the different methods used to make variable-star estimates. You've got more decisions to make so I'll provide you with enough information to help. When you first begin, try several of these methods. See which one works best for you. Then, when you feel comfortable, choose one. Then stick with it. As time passes, perhaps a few weeks, you'll find that your method becomes second nature and making estimates will become easier for you.

Estimating Magnitudes

With a little effort, you'll eventually find a variable star. When you've found one, you probably should estimate its brightness since this is at the heart of variable-star observing. Your accurate estimate is going to be important and you must take the time to do it properly. Initially, accurate magnitude estimates may require a little time to prepare. As you gain experience, estimates can be made rather quickly in most cases.

First, examine your star chart carefully. While using your binoculars or while at the telescope, identify the variable star and verify the relative brightness of all comparison stars. The first few times that you do this will require a little bit of time. After five or ten minutes you may start wondering if you're doing something wrong or if there is something that you don't understand. Don't be too concerned; this is common. Keeping track of your variable star and finding the comparison stars can be tricky at first. It won't be long before you begin to recognize the different star fields.

To estimate the magnitude of a variable star you simply determine which comparison stars are closest in brightness to the variable. Unless the variable is exactly the same brightness as one of the comparison stars, you will have to *interpolate* between a star that is brighter and a star that is fainter than the variable itself. This sounds simple and it is, but it takes practice to be skillful. Practice takes time and effort and all of this can only be done at night. These are reasons why you're the only variable-star observer in your neighborhood.

Your first few magnitude estimates need not be shared with anyone. Keep them a secret so you can side-step the pressure of making a near perfect estimate. Eventually, with a little practice, you'll develop more confidence in your estimates and then you'll feel good about reporting them. Let's take a look at how we actually make an estimate.

Estimating Magnitudes Using Interpolation

Any optical instrument's resolving power is greatest at its center of field, in other words, when the star is *centered* within the ocular. As a result, when the comparison star and the variable are widely separated, they should not be viewed simultaneously. They should be brought successively into the center of the field of view. You will do this by making slight adjustments, moving the optical tube assembly (OTA) of your telescope so that each star is centered. Some times you'll need to repeat this adjustment until you're confident with your estimate. Take your time doing this. It's important.

If the variable star and the comparison star are close together they can be placed at equal distance from the center. To do this, first mentally draw an imaginary line between the two stars. This imaginary line should be parallel to your viewing frame of reference to prevent what is known as *position angle error*. Turn your head or your erecting prism to improve this alignment. In the worst case, the position angle effect can produce inaccuracies of up to 0^m5 so it's important to eliminate this potential error.

I will mention the following several times because it is critical: *all observing must be done near the center of your eyepiece's field of view.* Usually, there is greater distortion of the image the further it is positioned away from the center of the field of view. Keeping the star centered, or nearly so, helps keep it within the distortion-free region of your eyepiece.

Use at least two comparison stars, and if possible, more. If the brightness interval between your comparison stars is very large, say 0^m5 or greater, take extreme care in estimating how the interval between the brighter comparison star and the variable compares with that between the variable and the fainter comparison star. Again, this is important.

Record exactly what you see, regardless of perceived discrepancies in your observations. You should go into each observing session with no expectations regarding the brightness of the variable star; do not let your estimates be prejudiced by your previous estimates or by what you think the star should be doing. This becomes more difficult as you tire during a long evening of observing. It's best to stop observing when you're very tired. Or at least, switch to casual observing and go look at the "pretty things."

If the variable is not seen because of extreme faintness, haze, or moonlight, then note the faintest comparison star visible in the region. If the faintest star visible should be $12^{m}\!.0$, then record your observation as < 12.0. The left pointing bracket is a symbol for "fainter than." This means that the variable is invisible and must have been fainter than magnitude $12^{m}\!.0$.

When observing variables that have an unmistakably red color, make your estimate using the so-called *quick glance* method rather than by prolonged stares. You remember that due to the *Perkinje effect*, red stars tend to excite the retina of the eye when observed for an extended period or time. So, red stars will appear to become excessively bright in comparison to blue stars if viewed too long. This can be a source of errors when estimating relative brightness.

Another technique for making magnitude estimates of red stars is called the *out-of-focus method*. Place the red star at the center of the field of view, then draw the eyepiece out of focus until it become visible as a colorless disk. The comparison star will also be out of focus and be visible as a colorless disk. Using this method, the systematic error due to the Perkinje effect can be avoided. If the color of the variable is visible even when the stars are out of focus, you may need to use a smaller telescope or an *aperture mask*. You may even need to consider using binoculars on bright stars. Another good reason to keep binoculars handy.

An aperture mask is simply a cover, with a hole cut into it, placed over the front of your telescope. The hole, smaller than the aperture of your telescope, should be cut off-center for Newtonian and Schmidt–Cassegrain telescopes to avoid the central mirror supports. For example, if you have an 8 inch aperture, you may find need for a 4 inch mask that reduces the amount of light entering your telescope. This mask can be made of sturdy cardboard or something similar. Simply tape it to the front of your telescope, taking care not to touch any lens or corrector plate. *Don't place tape on your lens or corrector plate!*

Using an aperture mask will change the focal ratio of your telescope. This isn't anything with which to really be concerned if you're visually observing. However, if you're using a camera it will be important to remember that your focal ratio has changed. Let's examine this idea for a moment.

Assume that your 8 inch telescope with a 2000 mm focal length has an *f*/10 focal ratio. This relationship can be shown by simply converting 8 inches to millimeters (approximately 200 mm); then divide the focal length of your telescope, in this case 2000 mm, by the aperture (2000/200 = 10). The answer is your focal ratio, in this case 10.

When you place your aperture mask on the front of the telescope, the aperture is changed from 8 inches to 4 inches. Now do the math. Your "new" aperture, 4 inches, equals approximately 100 mm, and your 2000 mm focal length remains the same but is now divided by 100. This gives you a focal ratio of *f*/20. Again, this is nothing to be concerned with when visually observing. However, when using a camera, it will become important.

For faint stars, you may wish to try making your estimate by using averted vision. To do this, keep the variable and the comparison stars near the center of the field of view while concentrating your gaze to one side. When doing this, you're using your peripheral vision. Move your eye around. If you are viewing at the extreme limit of your equipment, faint stars seem to "pop" into view. This is tough work. Sometimes, lightly tapping the side of your telescope will cause the star to come into view for a moment. Eye drops will help too. You can also place a large piece of black cloth over your head and eyepiece to block all ambient light. You'll notice that on some nights you can see fainter stars than on other nights. Be careful that you don't "wish" the star into view. A good rule of thumb is to see the star at least three times before recording it. Sometimes, it helps to leave a very faint star and come back to it in a few minutes.

The Argelander Method of Estimating Brightness

This is the method developed by Fredrich Wilhelm August Argelander in 1840. You must select two comparison stars that are close in brightness to the variable. Variable star charts have many comparison stars from which you may choose so doing so should be fairly easy. One of the comparison stars should be slightly brighter than the variable star and the other comparison star should be slightly fainter than the variable star. Using a method called *interpolation*, you're going to estimate the brightness of the variable with respect to the two comparison stars.

First, you will estimate the difference in brightness between the brighter star and the variable star. You will express the difference in brightness in steps. By convention, when using this method the brighter comparison star is designated "A" and the fainter comparison star is designated "B." The differences in brightness between steps is explained next.

- *One step*: At the exact moment of observation, if the brighter star (A) and the variable (V) seem to be equal, but after a moment of close examination the brighter star is *slightly* brighter than the variable, it is considered one step brighter and is recorded as: A(1)V. This means that the A star is one step brighter than the V star (variable).
- *Two steps:* If A and V appear equal when first observed, but almost instantly it becomes obvious that A is brighter than V, this is recorded as: A(2)V.
- *Three steps:* If a slight difference in brightness is obvious at the exact moment of observation, then A is three steps brighter than V and is recorded as: A(3)V.
- *Four steps:* If a distinct difference in brightness is immediately visible, this is considered four steps, recorded as: A(4)V.
- *Five steps:* A major difference in brightness between A and V is indicated as: A(5)V. You should be careful to choose comparison stars so that less than five steps are needed to make a good comparison. If you exceed five steps, this method rapidly loses accuracy. Good charts will help you select appropriate comparison stars.

After comparing the variable star with the brighter comparison star (A), use the same method to compare it with the fainter comparison star (B). For example, if the variable star is two steps brighter than the fainter star, it is recorded as: V(2)B. This means that the V star (variable) is two steps brighter than the B star.

After comparing the variable star with both the brighter and fainter comparison star, you will have a relation that may look something like this: A(3)V(2)B. This means that the A star is three steps brighter than the V star and that the V star is two steps brighter than the B star. Now you're ready to estimate the brightness of the variable star:

- *First*, determine the brightness of the A star. This will be on the chart. We'll say it's 11^m40 for this exercise. Remember, on star charts the decimal is omitted so that it won't be mistaken for a star.

- *Second*, determine the difference in magnitude between the bright comparison star and the faint comparison star. Let's say the fainter comparison star is 12^m30. On a star chart it will be labeled as "123."

- *Third*, divide the number of steps between the A star and the V star, in this case 3, by the total steps between all three stars. In this case the number of steps between the A star and the V stars is 3 and the number of steps between the V star and the B star is 2. The total number of steps is 3 + 2 = 5. So, we divide 3 by 5 (i.e. 3/5).

- *Fourth*, multiply the difference between the bright comparison star and the faint comparison star (11.4 − 12.3 = 0.9) by the number fraction that we determined in step three (3/5). That process will look like this (0.9 × (3/5)) and equals 0.54.

- *Fifth*, now just add the result calculated in step four to the bright star's magnitude (11.4 + 0.54 = 11.94). This is the estimated brightness of the variable star and should be rounded to 11^m9. Making comparisons to more than two stars will improve the accuracy of this method.

The important thing with this method is that there is no specific value attributed to any particular *step*. Each step is defined as the smallest difference in brightness that your eye is able to distinguish. The value of a step, expressed in magnitudes, will depend upon your local observing conditions and your experience. A beginner's step is often close to 0^m3; however, experienced observers may be able to distinguish steps as small as 0^m04. Compare your results with other variable-star observers. Don't be disappointed if your estimates don't exactly agree with other observers. With a little time, you estimates will be right in there with most other observers.

The Pogson Method of Estimating Brightness

Norman Pogson is a well-known variable-star observer from the nineteenth century. He developed a procedure that differs from Argelander's method in that each step is determined to be precisely $0^{m}.1$. This method requires you to compare a variable star with a single comparison star using a previously memorized interval of $0^{m}.1$. You then observe the variable again, using a different comparison star. The variable's magnitude is deduced later. Your first observation might be recorded as "A – 5," indicating that the variable star is five steps, or $0^{m}.5$, fainter than the brighter comparison star. Since you have already memorized what a tenth of a magnitude, or step, looks like, this observation is independent of the next, which considers the fainter star. In the second observation you might say "B + 4," meaning that the star is four steps brighter than the comparison star. Later, we would find out that A = 11.4, thus A – 5 = 11.9. If B were equal to 12.3, the B + 4 would also equal 11.9. Remember that a star gets fainter as its magnitude number gets higher.

The difficulty with this method is exactly memorizing the $0^{m}.1$ increments for your comparison stars.

Fractional Method of Estimating Brightness

With this method, you don't need predetermined magnitudes for the comparison stars, initially. You simply choose two stars, with "A" being brighter than "B", ensuring that the variable star's brightness is somewhere between these two comparison stars. Let's say that the variable is three-quarters of the way between the brightness of A and B. In other words, the variable is closer to the fainter star than the brighter star. You would record your estimate as "A(3)V(1)B."

Now let us reduce these estimates. We'll say that the A star is $9^{m}.80$ and that the B star is $10^{m}.30$ in this case. The difference between these two comparison stars is $0^{m}.5$. Now we divide 0.5 by the sum of 3 and 1 and the result is $0^{m}.125$. From the brighter star, we calculate:

$$9.8 + (3 \times 0.125) = 9.8 + 0.375 = 10^{m}.175$$

Or, using the fainter star, we calculate:

$$10.3 - (1 \times 0.125) = 10.3 - 0.125 = 10\overset{m}{.}175$$

Of course, we would round to $10\overset{m}{.}2$ since the accuracy of our comparison stars will limit the accuracy of our estimate.

These three methods will allow you to experiment with making variable-star estimates. Perhaps you may wish to use each method on the same variable-star observation, to see how they compare. Then, you may want to compare your estimates with other variable-star observers using VSNET, AAVSO or BAA VSS. In a short time, you'll become confident with one method and that will be the one that you use.

Let us now take a look at how estimating brightness is done when using instruments.

Photometry Capable Computer Programs

Several computer programs that will perform photometric measurements are available to amateur astronomers. Essentially, you load an image from your CCD, designate the variable star, as well at the comparison star and a check star. The computer program will then compare the brightness of each star. When finished, the program will provide a report indicating the difference in magnitude between each star. Programs such as this are absolutely required when using a CCD. When using a photometer, a spreadsheet can be developed to quantify your measurements.

Both of these methods are beyond the scope of this book but there are several books available to assist you should you begin CCD or PEP observations. A few that I would recommend are *The Handbook of Astronomical Image Processing*, by Richard Berry and James Burnell; *Photoelectric Photometry of Variable Stars*, by Douglas Hall and Russell Genet; and *Astronomical Photometry* by Arne Henden and Ronald Kaitchuck, all published by Willmann-Bell. Also, Arne tells me that he has a new CCD photometry book coming out after this one, so keep an eye open for it.

Collecting Data

As you are now aware, you have several choices of how you may observe variable stars. You may observe variables using your eyes and make visual comparisons with other stars that you can see through binoculars or the telescope's eyepiece; you may use a CCD and make comparisons with other stars in the image frame using computer software; and you may use a photoelectric method and make comparisons with other stars using mathematical interpolation.

Observing variable stars visually is the fastest method. It is really the only method available to amateurs who wish to make several dozen, or more, observations in one night and for observers who monitor cataclysmic variables (CVs) and wish only to detect the outburst. To detect a CV outburst, you need only use you eyes to detect the brightening of the star. When a cataclysmic variable goes into outburst it is evident and as long as your telescope has the ability to *see* deep enough, you will be able to detect these interesting outbursts. Also, to adequately monitor long-period, high-amplitude variables, such as the Mira variables, the precision of your eyes is completely satisfactory. You will normally make an observation of these slowly varying stars every two weeks or so and a change of several tenths of a magnitude will be quite evident to a careful observer. Visual observation is probably the fastest way to hunt for supernovae too. Don't think that you need to move on to instruments to completely enjoy variable-star observing. However, should you wish a new challenge, CCD and PEP methods await your attention.

CCD and PEP methods will allow you to observe and study variable stars with subtle variations, invisible to the naked eye, or to discover very fast, faint oscillations within a star's brightness, such as quasi-periodic oscillations (QPOs) in dwarf novae. The use of science filters is strongly encouraged when using instruments. Probably the most important thing about collecting instrument data, especially when filters are used, is that it must be able to be converted into some kind of standard data. In other words, when you use an instrument to collect data, for example a CCD or stellar photometer, the data produced is unique to that instrument and filter and it must be transformed into a universally accepted standard. Think of it this way; all humans will see the same event in a slightly different way just like all machines will detect the same event in a slightly different way. This is the result of imperfect manufacturing methods, non-uniform quality standards, perhaps even minor environmental conditions when the equipment is used and of course, observer induced inconsistencies will always be present to name a few.

Regardless of how you observe, each method will allow you to collect data in a slightly different way and one method should not be thought of as better than another. Each method has advantages over another as well as disadvantages. There is no room for snobbery within the variable-star observing community and no observer should feel compelled to defend their method of observing.

Searching for Supernovae

A hunt for supernovae is a little different from normal variable star observing. Instead of observing a local star, when hunting for supernovae you're going to be exploring distant galaxies so a slightly different technique is used. When searching for supernovae, the galaxies in which you search need to be chosen according to their distance. A common mistake when hunting supernovae is to choose galaxies according to their brightness and thereby assume the brightest galaxies are the closest galaxies. While this can be correct in many cases, there are many near-by, less luminous galaxies as well as distant, bright galaxies.

Another concern for supernova hunters is that galaxies need to be chosen according to their type. Referring backing to Chapter 5 you'll recall that type I supernovae require a white dwarf with a companion star. Because white dwarfs are old stars you can deduce that type I supernovae appear only in the older star populations, for example within the central bulge of a spiral galaxy and of course within the old star populations of elliptical galaxies. However, type II supernovae are the result of massive, quickly evolving young stars and therefore appear only within galaxies where stellar formation is still underway, for example the younger spiral galaxies.

When conducting a supernova hunt, some homework is needed before beginning. A little research will maximize your search method and improve your chances of discovering one of these rare events.

Recording Data

The following information should be recorded within your log book as soon as possible after each observation:

- name and designation of the variable
- date and time of your observation
- magnitude estimate for the variable
- magnitudes of the comparison stars used for the estimate
- identification of chart and scale used
- notes on any conditions which might effect seeing.

I've read about observers who memorize dozens, even hundreds, of observations from a night of observation, then record their information later or even the next day. I recommend that you record your observations as you make them. Develop good habits now so that you don't have to break bad habits later.

After recording your observations, you may then wish to report your variable-star estimates to one of the recording organization such as VSNET, BAA VSS, or the AAVSO. You'll need to report your estimates in the proper format. Each organization uses a slightly different format and it's important to use the correct one. They're usually straightforward and simple. Visit each organization's Web site and become familiar with their reporting procedures. See Chapter 14 for a list of some of the organizations and their Web sites.

Reporting New Discoveries

Occasionally you will make a discovery. After recording your data, if you think that you have made a discovery it is important that you report your suspicions accurately and quickly. It is also important not to waste other people's time with false claims.

First, if after rigorous examination you are convinced that you have made a discovery, it is important to determine its position as accurately as possible. You may need assistance doing this. Reporting a suspected discovery by saying that it's a little to the left, and down, from the big bright star, won't work. You'll need to determine its right ascension and declination to a fairly good accuracy; within a couple of arcsec, at least!

Second, you must ensure that you have a bona fide discovery! Take extraordinary care when doing so. You *must* get a second opinion! Call a friend, or use the Internet to request assistance. Your discovery will remain secure if it is in fact a real discovery but getting other astronomers to help determine it's existence is important. Calmly request help. You'll find many observers willing to help you.

It may take a day or more for your discovery to be recognized so be patient. Many discoveries turn out to be something that is known but not well observed, so be prepared for this too. In time, undoubtedly you'll end up observing something that you can call a discovery so hang in there.

Chapter 13

Variable Star Data Management, Reduction and Analysis

> There are three kinds of lies: lies, damned lies, and statistics.
>
> *Benjamin Disraeli*

"What kind of story is my observations telling me?"

Without a doubt, your observations of variable stars will provide some insight into the specific characteristics of each. Of course, the quality of your observations will determine the depth of the insight. In any case, before long, you will find yourself the custodian of a large amount of information that may have scientific value. It will certainly be valuable to you. Your ability to properly analyze the growing mound of information and to tease from within the hidden secrets will increase its value. With an understanding of basic data analysis, interpreting the secrets hidden within your own observations can be rewarding and is an important part of variable-star observing. Basic analysis of your data will elevate you above simple bean counting.

As a result of your VSO activities, presumably you will soon begin to suffer from information overload. Consider observing only 20 or 30 stars, once a week, over the course of a year. This modest schedule will result in over 1500 observations. Observe a greater number of stars or observe more frequently and this number will increase dramatically. Unquestionably, some observers never return to their past observations once they are collected and reported. Some observers are interested only in observing and reporting. The story contained within their observations remains unread except by others.

However, many variable-star observers enjoy conducting long baseline time analysis of their program stars and then comparing their observations from past seasons in an effort to confirm a predicted characteristic or detect anomalous behavior. Basic analysis of your data can be fascinating when examining the mass transfer between entangled eclipsing binary stars, RV Tauri stars exhibiting the interesting RVb phenomenon, RR Lyrae stars demonstrating the Blazhko effect, long-period Mira variables displaying fluctuating periods and amplitudes in response to internal stellar dynamics or when following superhump evolution within SU UM*a* stars. These are a few of the interesting phenomena that you can study and there are many more interesting characteristics of variable stars that basic analysis will expose.

Many visual observers dedicate two or three hours, a couple of nights a week, to observing, particularly during the winter months when darkness comes early, the air is crystal clear, and the stars intensely sharp. During exceptional evenings, you'll be able to make a visual estimate every minute or two (taking into account bathroom breaks, cups of hot cocoa, talking with colleagues, recording your estimates, roaming the star fields, etc.). A great evening of observing can easily produce over a hundred visual estimates. Using a CCD and the automatic capabilities of many telescopes available today, in a short 8-hour night it's relatively easy to collect 500 digital images or more. Either way, after a handful of nights of observing you won't have enough daytime hours to properly analyze your data. You'll begin to wish for a cloudy night or two just so you can catch up. In extreme cases, you'll just have to quit your job!

At this modest pace, in a few years, you may find yourself struggling to find a handful of observations that you made a season or two ago among the many thousands that you've made and recorded over time. As a visual observer, you may possess books of data and if you're using a CCD, thousands of images will eventually fill your hard drive. With a stellar photometer, since you will need to make sky measurements for your comparison star and check star, as well as your variable star, your resulting records will grow at an even faster pace. How are you going to manage what will eventually become a huge amount of information in such a way that you will be able to retrieve individual observations when needed?

At this point, you're wondering, "Why do I need old observations anyway?"

Well, old observations will be needed to compare with newer ones when searching for suspected period changes and amplitude changes. Or when verifying comparison and check stars from a season or two ago. Checking previous integration time, camera temperature and which filters were used can be important for CCD and PEP observers. Perhaps, a request to share your data with another observer will arrive. Or even checking a star field in an older CCD image because a recent image shows a "new" star. Past observational data is valuable. By virtue of some weird universal law, your observations are most valuable when you can't find them. Suffice it to say, finding data from past observing seasons can be frustrating without a formal storage and retrieval method.

Database Management

In most cases, you will be able to use a computer database manager or even a spreadsheet to organize your information. A simple relational database, such as *Microsoft Access*, may provide satisfactory results. Many database programs exist and I mention *Access* only because that is what I use.

A database is a collection of information, such as variable-star estimates, and a database management system (DBMS) is a tool designed to help you manage the information in your database. A table of information neatly organized into rows and columns is called a relation, and a relational database management system is one that is specifically designed to manage information that's organized into one or more tables. Usually, variable-star data is best organized into a table format and as a result, it is fairly easy to store within a relational database.

I'll use my method as an example. It's not perfect, it may not be for everyone and I'm sure that not everyone will like it but it's just an example. You will no doubt develop your own method that will continue to evolve through the years. I use a relational database with just a few field names. The field names are the categories containing information by which you want to sort. For

Table 13.1.

Object/description		Method	Date	Data pts	Location of data
R Vir	Mira variable	V	April 1, 2000	1	Obs Log, Apr 1, 2000, page 147
S Vir	Mira variable	V	April 1, 2000	1	Obs log, Apr 1, 2000, page 148
W Vir	W Vir variable	C	April 1, 2001	6	CCD images (WVir), ZIP Disk #31
ST Vir	RR Lyr variable	P	April 1, 2001	120	Obs log, Apr 1, 2000, page 149–53

example, you may want to sort by object to see all of the entries you've made for a particular variable star or by date so that you can find all of the data from one particular date. Maybe you'll wish to sort by variable star type to see how many different classes of variable stars you have been observing. Perhaps you will want to sort by type of data, in other words, whether it's visual, CCD or PEP data. An example of the database table is shown in Table 13.1.

Using this simple method, I can find a particular variable-star estimate or PEP record within my personal observation log or locate one, or more, CCD images stored on a computer ZIP disk or CD. I keep a detailed personal log so in most cases I simply refer to a particular date to find the observational information for a particular star. When looking for data that spans months or years, a *database sort* will tell me which observation logs I need to get and which date and page to find. When I need to find a CCD image that is stored on a ZIP disk or CD, for example, the database entry tells me which disk holds the image or images.

With this uncomplicated method, I can sort my fields by object to find all of my observations made of one particular star or by date, to look at my activity on a particular day. I can sort by variable-star type, to see what kinds of stars I've been observing and I can also sort by the method that I've used to obtain my data: visual, CCD or PEP.

When I store my CCD images, I label each digital image using the brightest variable star found on the image, usually the target of my interest. In some cases, the variable in which I'm most interested is not the brightest variable found within the CCD image but I still name the image after the brightest variable star. This method is similar to how AAVSO labels their star charts: using the brightest variable star even though many other variables star can be found on a chart. When I store CCD images that are without a variable star, I use a catalog name for the brightest star as the image label. Galaxy images, used for supernovae hunts, are named by NGC or other catalog name.

Individual CCD images are named using the brightest variable star found on the image, followed by the date, then run number, filter type followed by sequence number. For example, a group of six images of W Vir taken on April 1, 2000, during the first run of the evening, using a V-filter, will be labeled: WVIR20000401-1V_000, WVIR20000401-1V_001, WVIR20000401-1V_002, WVIR20000401-1V_003, WVIR20000401-1V_004, WVIR 20000401-1V_005. Remember, astronomers use the number zero when recording sequences. Start with zero, not the number one. It will keep you consistent with other astronomers. This method allows me to sort my images by object, date, run, filter type and sequence.

I also keep a worksheet, prepared using a spread-sheet, for each star that I observe. By doing so, I can quickly check all of my observations, spanning years, and prepare a quick light curve or phase diagram. My worksheet includes the star's name, variable class, date/time of observation, estimated magnitude (when visual) or differential magnitude when using instruments, comparison and check star identification and any relevant notes concerning weather, wind, moon or anything else that I feel is important. When taking PEP measurements, I record the sky background brightness too. This may sound like a lot of work but it isn't. Once it becomes habit, it's easy, quick and you'll really appreciate whatever system you use once you start searching for old observations. I'd guess it takes a few seconds to enter the information for each observation. Much of it is redundant, such as date or star name so simply using a *copy and paste* operation on my computer eliminates making the same entry more than once. Customizing a menu specifically for your astronomy needs and the use of macros helps too. Programming your computer system to anticipate your needs is a great time saver. The real value of all of this is that I can find any observation or sequence of observations, some made years ago, in about 30 seconds.

Develop you own method with your own require-ments in-mind, but try and develop your system within the first year of use. After you've been using one method for about a year, changing it is a real pain!

Data Reduction

After collecting your observational data, the first step that you must take in analyzing variable-star data is to organize it. When organizing variable-star data, the date/time of the observation, the estimated magnitude of the variable star and the star with which it is being compared and perhaps a remark regarding special conditions are necessary. In some cases, more information is recorded but for basic analysis of variable-star data, we will use the date/time and magnitude of the star being measured. Try to configure your database so that your entries match your analysis needs. By considering this, you can use copy and paste operations to move blocks of data quickly.

Here is an example of a journal entry made by a variable-star observer:

Variable Star Journal

Date	Star	Time	Est. Mag.	Notes
Sep 27, 2001	R Vir	9:30 PM	10.1	Clear with moon
Chart xxxx, comparison stars 10.5 and 9.8				
Sep 27, 2001	R Per	9:33 PM	12.1	Clear with moon
Chart xxxx, comparison stars 12.5 and 11.5				
Sep 27, 2001	R Psc	9:37 PM	9.2	Clear with moon
Chart xxxx, comparison stars 9.7 and 8.8				
Sep 27, 2001	R Ari	9:41 PM	11.4	Cloudy with moon
Chart xxxx, comparison stars 11.9 and 10.9				

As you can see, something simple is all that is really needed. Here the observer records the star's name, the date and time of the observation, the estimated magnitude and the charts/comparison stars used. Try this simple method first. You may not need any more information that this. Don't allow record keeping to become a burden so that it interferes with your observing.

After a night of observing you should have a record of your work. You're next effort will probably be in reducing your data. Reducing your data is simple converting it into a form that allows proper analysis. In most cases, you'll not be required to do much with your estimates before you begin to conduct a proper analysis of your data. However, in situations where you are using instruments, you're reduction procedures can be a bit time-consuming. It would be impossible to describe all of the procedures within this book. It really requires a more in-depth description, so I recommend a couple of books specifically written for CCD and PEP observers. Two good books are *Photoelectric Photometry*, by A. Henden and R. Kaitchuck and *Photoelectric Photometry of Variable Stars*, by D. Hall and R. Genet, both published by Willmann-Bell. Both provide excellent descriptions of data reduction for CCD and PEP users.

If you're going to simply report your data and conduct no analysis yourself, all that is required is to put your observations into the proper format. Each reporting organization, such as VSNET, AAVSO or BAA VSS, has their own format. Visit their Web site and see what their wishes are in regards to reporting format.

Analysis

If you've decided to conduct some analysis of your data, it's time to read the story the stars have been trying to tell you. Statistics will be the tool that you will use to understand the story. Generally, the study of statistics is divided into two parts: descriptive and inferential. *Descriptive statistics* describes number sets, such as how many numbers, the middle number, and the spread of the numbers. *Inferential statistics* uses some of these descriptions to make inferences or guesses about an entire population based upon a sample of the data. You will use both to analyze the data that you produce from observing variable stars.

Two types of information can be collected in statistical studies: *qualitative* and *quantitative* data. Quantitative data consists of measures or quantities that can be put into an order or ranked in some way, such as estimates of star brightness or time between outbursts. Qualitative data is not comparable by arithmetic relations; for example, how much fun you're having observing variable stars or how tired you feel at 3 o'clock in the morning.

An additional distinction exists among quantitative data. Quantitative data is defined as either *discrete* or *continuous* data. Discrete data is measured in exact, or discrete, numbers. Data that can assume any value within an interval, or between two numbers, is an example of continuous data. Individual variable-star magnitude estimates are considered discrete data and each individual estimate that you make will consist of one discrete number. However, variable-star brightness

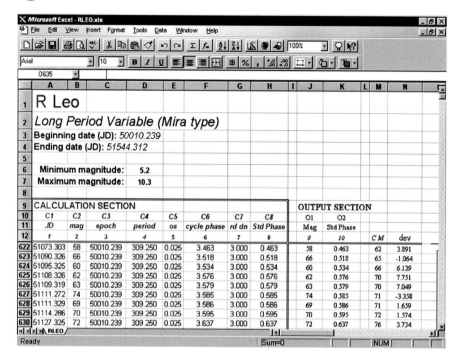

Figure 13.1.
Observation data for the
Mira-type variable, R
Leo. Data provided by
VSNET.

can assume any number, between a maximum and a minimum, and is therefore considered continuous.

Your goal as a variable-star observer is to estimate the brightness of a star, at a particular time, and describe that estimate as an accurate discrete measurement. Later, by using many accurate discrete measurements taken over a period of time, you will be able to construct a model of the continuous data that represents the star's variations over time. In other words, you will be able to develop a predictive model of some characteristic of the star, in this case the variability of the brightness of the star, and prediction is one of the fundamental goals of science. It's also fun.

Take a close look at Figure 13.1. The data shown is provided by the *Variable Star Network* (VSNET) and relates to the star R Leo, a Mira-type variable. The first column is the Julian date (2,400,000 +). The second column contains the magnitude estimates.

This arrangement is an example of the data collected by a variable-star observer or a group of observers. The date of each observation is a Julian date but you will notice that it has been truncated or shortened. It has been shortened by recording only the last five date digits. The full Julian date for the beginning of this data

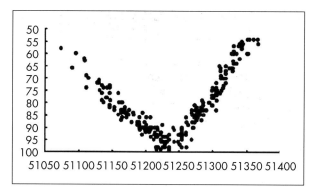

Figure 13.2. Light curve of R Leo produced from the observation data. Data provided by VSNET.

is 2451,073.303. It's a common practice among variable-star observers to shorten the Julian date and use only the last five or four numbers of the date. Care must be taken when using old data since the last four digits of a Julian date will eventually recur in a relatively short period of time. To determine the actual Julian date for any of the records shown here, just add 2400,000 to the date. The decimal portion of the date is the time element.

This table of data may look interesting and ordered, but reading the story that is concealed within will require a bit of work on your part. You're going to need to use a different language to read *this* story than you are using to read this book. Now is when the language of mathematics, specifically statistics, becomes impor-tant.

If we were to graph the data[1] provided in Figure 13.1, we would be able to see each data point representing a single estimate of magnitude (*star brightness*) placed in chronological sequence (*date of observation*). You can see that over a time period of about 300 days, this star has varied in brightness by a little less than five magnitudes (Figure 13.2).

Supposing that the data is correct, it's obvious that this star is varying in brightness. That piece of information is valuable in its own right but there is a deeper, more interesting story here. Before we learn to read the story, let's look at some of the things that hide or distort the real story that you are trying to read.

---

[1]To learn how to construct a graph similar to the one shown, refer to Appendix C, "Spreadsheets using Microsoft Excel".

Errors

As with any story, there are going to be some inaccuracies, a few errors. Some of these errors are going to be your fault and some will occur for reasons beyond your ability to control. In fact, if you're not mindful, you may not even know that errors are creeping into your observations. It's important to eliminate those errors that are your fault and take steps to identify those that are beyond your ability to control so that you can properly consider them. Don't be offended; everyone has errors, even professional astronomers. You just have to account for them.

The first step that you can take to reduce errors is to pay attention. Be alert and take yourself, and your work, seriously. Formalize your observation and analysis process so that you follow the same procedures every time; notes, a worksheet or a checklist will help. Also, wait until you're rested to conduct detailed analysis of your data. Accuracy is more important than speed!

Errors that you will not be able to totally control but that you might be able to influence in some small way include atmospheric extinction and environmental conditions. Atmospheric extinction is caused by making estimates through a thick atmosphere, such as when you observe a star positioned very low near the horizon. When a star is low, you're looking through a lot of air. This air, the atmosphere, is going to change the light that reaches your eyes or instruments. It's best to observe stars when they are high overhead. Think of the atmosphere as an ocean, an ocean of air and you are sitting on the bottom of this ocean looking up through it.

Of course, there will be exceptions, such as stars that never get very high relative to your location. Proper planning will help here. Plan to observe stars when they are highest in the sky. The *meridian*, an imaginary line running from south to north (or north to south, depending upon your perspective and hemisphere), will help you. When a star passes through the meridian it reaches it's highest point in the sky. This event, moving across the meridian, is called the *transit*. Observing stars when they are close to the transit will ensure that you are viewing them when they are highest in the sky. This is most important for those stars that never get very high in the sky.

Environmental conditions include the weather, site location and the Moon and Sun. For example, try not to make observations through light clouds or if possible, from heavily light-polluted city locations. Be careful when observing stars close to a bright Moon. Making observations just as the Sun sets or just before it rises can influence your estimates too. Again, there will be exceptions. Strive for the best observing location, best conditions and best orientation relative to the star that you are observing. When your observations are influenced in some way that may effect their accuracy, make sure that you indicate it on your records.

Rejection of Data

During some part of the story that you are recording with your observations, you're going to begin to doubt your data for reasons that are impossible to predict here and now. You may come to believe that your observations should not be used, reported, or that they should even be thrown out.

Resist the temptation to throw out observations! Keep all of your observations. Remember, you don't need to report those for which you've lost confidence. It may be that some type of error has occurred and that your observations are wrong. It could also be that something extraordinary has occurred and you've accurately recorded it. When you have reason to question your data, indicate it in your log book but don't throw observations away. As with most things in life, you can learn from your mistakes when observing variable stars.

If you have truly made an error in your observations, try and figure out what you did wrong. Look at your estimates. Think about what you did. Review your log. If you can find your mistake, you can avoid it in the future. Otherwise, you may just make the same mistake again.

Spreadsheets

A spreadsheet is an excellent tool to analyze your variable-star data. Get to know the capabilities of your spreadsheet and how your observations can best be displayed. Charts and graphs allow you to visualize your data and make it much easier to understand. A column of numbers may look neat and organized but a scatter diagram can tell you an interesting story.

Spreadsheets allow you to display your information in many ways ranging from two-dimensional line graphs to three-dimensional contour maps. Selecting the proper display is important. If you've used spreadsheets before, and if you're familiar with statistics, then you know that you can support or refute almost any position using the same data. Because this is possible, you must be careful. Use methods that are familiar and accepted. If you're not sure what they are, ask. Present you observations with an honest intent to display your observations accurately.

With all of that said, you're probably wondering what you can do with your data and a spreadsheet. You're going to be very pleased with the possibilities! Now is when those lonely, cold nights of observing will result in something that you can show someone. You will be able to quantify you work and produce something that you can touch and see; something more than a number written on a page.

Light Curves

Light curves allow you to plot the changing brightness of a variable star over time. All that is required is to put the date and time (the Julian date) of your observations into one column and place your brightness estimate into another column. Make sure that you match the correct estimate with the proper date and time.

Usually, you'll have several choices regarding the type of chart you wish to use, such as a line, point or bar. In most cases, a point chart is best. This type of chart displays each brightness estimate as a point correctly correlated with the proper date and time. Within spreadsheet programs, these charts are sometimes called XY (scatter) charts.

Since your desire is to show a trend over a time interval, it's important to decide how you want your chart to look. Usually, the time interval is displayed along the *horizontal axis* (X-axis) and the brightness estimate is displayed along the *vertical axis* (Y-axis). Also, since you are going to compare observations made at uneven intervals (even a few seconds difference between observations can be important), the XY (scatter) chart is best. This chart plots data in which the *independent variable* (time) is recorded at uneven intervals. Your brightness estimate is the *dependent variable* because the actual brightness of the variable star depends upon when you make the estimate.

Light curves allow you to analyze the gross variations in brightness that a variable star is experiencing. You can see how a star varies over the course of seconds, minutes, hours, days, a month, a year or over the course of decades. Now it's time to learn how we can analyze the subtle variations experienced by some variable stars.

Phase Diagrams

When the same cycle repeats over and over, it is referred to as periodic behavior but not all variable stars are periodic. If the variable star is periodic and you want to know what is happening at any moment during a cycle, it doesn't matter which cycle you're observing because every cycle is exactly the same. What does matter is which part of the cycle you're observing. So if a star is strictly periodic, then its variation depends only on where it is in its cycle, a quantity called *phase*.

Phase is measured in cycles and because phase is measured in cycles, a single cycle starts at 0 and ends at 1. A phase of 0.5 corresponds to 50% through the cycle. A phase of 0.2 corresponds to 20% through the cycle and a phase of 1 corresponds to 100% through the cycle. After a cycle has reached 100% of its phase it begins anew.

To compute the phase in terms of cycles, you need to know the length of each cycle in seconds, minutes, hours, or days. In other words you need to know the *period*. In this book, to compute the period of a variable you'll use the computer program TS11. We'll discuss TS11 in a few moments. I'll tell you were to get a free copy and you can begin using it immediately.

To compute the phase in terms of cycles you also need to know the starting time of the cycle, called the *epoch*. These may be new terms to you but don't let them intimidate you. You'll be using them often so they will become familiar in a short time.

These two quantities, the period and the epoch, will enable you to compute the phase at any given time.

Suppose the epoch is at t_0 (time at zero) and the period is P and you wish to calculate the phase at some time t. First we find how far we are into the cycle, by simply subtracting the starting time:

$$t - t_0$$

For example, if t_0 is JD 4500 and t is JD 4600, you are 100 days into the cycle. To calculate the phase in units of cycles, you simply divide this quantity $(t - t_0)$ by the period:

$$\phi = (t - t_0)/P$$

The symbol ϕ is the Greek letter *phi* and it is used to represent the phase, in cycles. Let's say the period is 500 days. If you divide 100 by 500 you will see that we are 20% into the cycle or at phase 0.2. When you compute a phase, you're interested in what is called a *standard phase*. To calculate the standard phase, you're going to need to know the period of the variable star.

Computer Programs for Data Analysis

Available today are computer programs that will assist you in performing sophisticated mathematical analysis of your data such as searching for and fitting sinusoidal patterns within time series data, Discrete Fourier analysis, and phase dispersion minimization (PDM). Without possessing an advanced degree in mathematics, you will be able to use these powerful mathematical tools to aid you in understanding variable stars. Of course, an understanding of mathematical methods that goes beyond the elementary level will enhance your understanding and appreciation of your analysis. This deeper understanding of the necessary math will come with a little effort once you start using these various methods and after you feel a need to really understand what your data is saying to you. Don't let the math intimidate you.

In all likelihood, one of your first needs for sophisticated computer analysis will be to determine the period of a variable star. There are several things to remember regarding the search for a variable star's period: not all variable stars are strictly periodic (in other words, there may not be a detectable period),

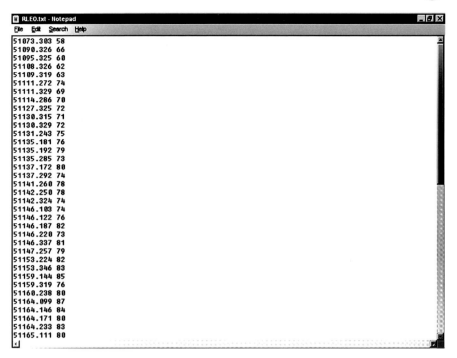

Figure 13.3. R Leo data organized for TS application. Notice *one space* between the JD and magnitude estimate. Data provided by VSNET.

many variable stars have more than one period, and there are going to be gaps in your data.

A fast and simple program for detecting a variable star's period that is available from the AAVSO is TS11 (TS). This program is available at their Web site (*www. aavso.org*) and is free. There are other programs available, some more complex than TS11, but we'll use the AAVSO program here. When you feel the need for more advanced programs, move on. It's not unusual to use more than one or two programs to conduct period analysis.

TS needs two pieces of information: time and magnitude. The program expects your data file to be a text file (*name.txt*) or data file (*name.dat*) and to consist of columns with **one blank space** separating the columns. For example, the data from Figure 13.1 was stored as "RLEO.txt" and consisted of the two columns shown (Figure 13.3). The maximum number of data points that can be loaded at one time is 4000. If you wish to analyze more than 4000 data points you must do so using additional runs.

During the analysis run, TS stores the processed data in a second file (*name.ts*) that you can further analyze using a spreadsheet. We'll use *Microsoft Excel* in this

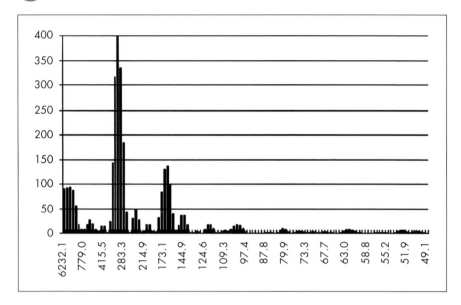

book but any spreadsheet will provide the basic tools necessary for fundamental analysis.

TS gives you the ability to perform the following mathematical operations: average your data, fit a polynomial (using the method of least squares), compute residuals, discrete Fourier analysis (for period and frequency), and model your data. The instructions for TS are included in the files that you download from AAVSO. Plan on spending an hour learning how to use this program.

Figure 13.4 shows the power spectrum, illustrated as a two-dimensional graph, that displays the Fourier analysis of the data from Figure 13.3.

This chart was made using the file *RLEO.ts* produced by the program (Figure 13.5).

Frequencies (displayed along the X-axis) indicated by the highest power levels (displayed along the Y-axis) are the most likely candidates for the actual frequency of a periodic fluctuation in your data. A critical examination of the tabular data stored in the file (*name.ts*) produced by TS allows you to precisely select the best candidate frequency. The chart is an imprecise tool that allows you to visualize your analysis so you must carefully analyze the *name.ts* file. Nothing will replace hands-on experience for becoming familiar with this program.

You can see that the frequency of 0.003209166 (the period equals 24 divided by the frequency and then divided again by 24, in this case, 311 days) corresponds

Figure 13.4. Power spectrum produced using TS-11 using R Leo data. Data provided by VSNET.

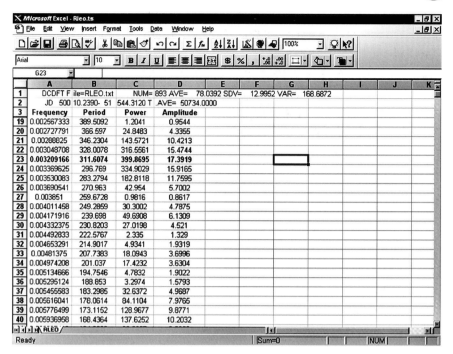

	A	B	C	D	E	F	G	H	I	J	K
1	DCDFT F	ile=RLEO.txt	NUM=	893 AVE=	78.0392 SDV=	12.9952 VAR=	168.6872				
2	JD 500	10.2390- 51	544.3120 T	.AVE=	50734.0000						
3	Frequency	Period	Power	Amplitude							
19	0.002567333	389.5092	1.2041	0.9544							
20	0.002727791	366.597	24.8483	4.3355							
21	0.00288825	346.2304	143.5721	10.4213							
22	0.003048708	328.0078	316.5561	15.4744							
23	0.003209166	311.6074	399.8695	17.3919							
24	0.003369625	296.769	334.9029	15.9165							
25	0.003530083	283.2794	182.8118	11.7595							
26	0.003690541	270.963	42.954	5.7002							
27	0.003851	259.6728	0.9816	0.8617							
28	0.004011458	249.2859	30.3002	4.7875							
29	0.004171916	239.698	49.6908	6.1309							
30	0.004332375	230.8203	27.0198	4.521							
31	0.004492833	222.5767	2.335	1.329							
32	0.004653291	214.9017	4.9341	1.9319							
33	0.00481375	207.7383	18.0943	3.6996							
34	0.004974208	201.037	17.4232	3.6304							
35	0.005134666	194.7546	4.7832	1.9022							
36	0.005295124	188.853	3.2974	1.5793							
37	0.005455583	183.2985	32.6372	4.9687							
38	0.005616041	178.0614	84.1104	7.9765							
39	0.005776499	173.1152	128.9677	9.8771							
40	0.005936958	168.4364	137.6252	10.2032							

Figure 13.5. Time series data file produced from analyzing R Leo data.

to the highest power rating (amplitude) of 399.8695. This power rating is a numerical discriminator. Of course, to be absolutely sure, further analysis will be necessary before you can safely conclude that the highest power rating relates to an accurate period or frequency. Occasionally, the highest power rating *does not* propose the correct period or frequency. Check your work. Then check it again.

This data was checked for a period every 12 hours. If this particular star, as a result of its intrinsic behavior, possessed a period of less than 12 hours, we would miss it using this resolution (for example, a δ Scuti type variable). On the other hand, searching with a resolution of 0.0001 day over a period of 300 days will take a significant amount of computer time that could be better spent. This is where your understanding of stellar evolution and variable stars becomes important. You should have, at least, an educated suspicion regarding the variability of the star or stars that you will study. Occasionally, you will have no idea but these types of situations should be relatively rare. Good preliminary work on your part should reduce wasted time and improve your ability to conduct good analysis. Check the literature and look at other observation reports.

Examples of the questions that you will be asking yourself may approximate the following:

- What resolution is best based upon the stellar spectral type?
- Is there a possibility of multiple periods? Is there a possibility that the star in not strictly periodic?
- Is my data baseline too short to detect long periods?

The Fourier analysis just discussed is a good tool for detecting and quantifying periodic fluctuations in times series if by periodic we mean of truly constant period, amplitude, and phase. Real astrophysical systems rarely exhibit such consistency of fluctuation. Often periodic fluctuations arise intermittently as transient phenomena. Even for a time series with consistent periodicity you will usually see time evolution of the parameters of the fluctuation. Discrete Fourier analysis can detect, and to some degree quantify, such behavior, but it is far from ideal for such purposes.

So, "What is Fourier analysis?"

Jean Baptiste Joseph, Baron de Fourier was born on March 21, 1768. A physicist, Fourier was studying how heat flows through an object when it is heated. He was able to determine that the movement of heat also behaves like a wave. After careful study, Fourier discovered that, although in a very complicated way, heat waves are periodic waves. Periodic waves consist of the same pattern or waveform repeated over and over and Fourier discovered that no matter how complicated, a wave that is periodic with a pattern that repeats itself, consists of the sum of many simple waves. His method is known as Fourier analysis.

Fourier analysis is a powerful tool used by variable-star observers to determine the period exhibited by some variable stars. Of course, the important consideration is periodic behavior. The variable star to be analyzed *must* be periodic or at least approximately periodic. As you know by now, not all variable stars are

Figure 13.6. Light curve of AM Her showing the nonperiodic nature of the outbursts. The Julian date is indicated along the horizontal axis.

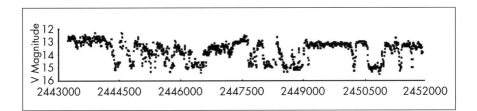

periodic. For example, cataclysmic variable stars such as dwarf novae are not periodic. Their period, the time between outbursts, is an approximate average. For example, a light curve of the cataclysmic variable AM Her is shown in Figure 13.6.

Chapter 14
Variable Star Reporting and Recording Organizations

It is therefore necessary that memorable things should be committed to writing ...

Sir Edward Coke

"Each observation is unique and can never be repeated."

This statement expresses the underlying significance of each observation that you make. However, it is not only the actual observation that is unique but also your personal experience each night. Every night of observing *will* be a memorable affair. You may be with friends or a family member, you may be observing at a new location, you may witness in solitude, a meteor's fall from the heavens or an aurora resulting from solar activity two days earlier. Each of these different situations, and myriad others, will define many nights of observing that you will want to remember. Don't lose something important because you didn't record its happening. Make it a habit to write down in you log or record book everything that is not mundane. You won't regret it.

By using the Internet, it is possible to immediately communicate with, and report your variable-star observations to individuals, groups and organizations from around the world. One of the most valuable benefits of using the Internet will be your ability to compare your observations with others to see how well you are doing when estimating brightness or to look at estimates made during times that you are unable to observe.

Included here are a few organizations that provide assistance, guidance, publications and support to amateur variable-star observers. Certainly there are more organizations than can be listed in this chapter and I have not omitted any organization intentionally. One of the joys of amateur astronomy is discovering the treasures hidden within the Universe, including the Internet.

Contact one or more of these organizations once you feel comfortable with observing variables stars; perhaps after a week or so. At least, become familiar with some of the reporting procedures. You need not wait until you feel completely competent but understanding the basic methods, nomenclature and various other aspects of the hobby will make communicating a bit easier. You'll feel less intimidated too.

I can't recommend one organization over another; they're all good and all will welcome you as a participant or as a member. Don't worry about the number of observations you've made (or haven't made), the equipment you have (or don't have), or the length of time you've been observing variable stars. Within each organization you'll find many observers, new and old, who will have questions and concerns similar to yours. Check them out. They not only want your data, their membership is composed of hundreds of other variable-star observers. Each began their study of variable stars much as you have started yours. You will find most members more than willing to help you, answer your questions, share their ideas and assist you in many ways.

British Astronomical Association, Variable Star Section (BAA VSS), Burlington House, Piccadilly, London, W1V 9AG *(http://www.telf-ast.demon.co.uk/)* The British Astronomical Association (BAA) was formed in 1890 and the Variable Star Section (VSS) was created the following year with the goal of collecting and analyzing variable-star observations.

Observations are reported through *The Astronomer*, a monthly magazine publishing new observations in all fields of astronomy. As well as assisting amateur astronomers monitor the activity of hundreds of variables and develop light curves, their database enables the section to supply records to professional and amateur astronomers for analysis. An excellent organization composed of very serious amateur astronomers.

The Variable Star Network, Kyoto University (VSNET), Kitashirakawa-Oiwake-cho, Sakyo-ku, Kyoto 606-8502, Japan (*http://www.kusastro.kyoto-u.ac.jp/vsnet/*) The Variable Star Network (VSNET) should be recognized as one of the best resources for amateur astronomers when it comes to the study of variable stars simply because of volume. This group of professional astronomers produce nightly lists of variable-star observation reports, monthly activity reports, alerts, and calls for collaboration as well as answering question asked by the amateur astronomy community. The VSNET is found within the Department of Astronomy, Kyoto University, and consists of many, many mailing lists that are used to distribute various messages pertaining to variable stars, especially cataclysmic variables (CVs) and related objects. VSNET distributes observational data regarding the discovery of supernovae, novae, rare outbursts, discovery of new variable stars, and dramatic changes of known variable stars, including compiled data and preprints. VSNET also provides discussions more diversely related to variable stars and variable star observing as well as providing finding charts. They also offer Mira and eclipsing binary data exchange and discussions, newly discovered variable stars and VSNET circulars for CVs and long-period variables. Another excellent organization.

The **Astronomical Society of South Australia** (ASSA), Honorary Secretary, Astronomical Society of South Australia Inc, GPO Box 199, Adelaide, SA 5001, Australia (*http://www.assa.org.au/info/*) The Astronomical Society of South Australia was founded in 1892 and is the oldest society of its kind in Australia. It is the only representative body for amateur astronomy in the state of South Australia. Membership is open to people of all ages and professions – the only prerequisite is an interest in astronomy. The objectives of the Society are to promote the science of astronomy and all its branches. You will find charts and information regarding many southern hemisphere stars here. Very serious amateur astronomers with excellent information.

The **American Association of Variable Star Observers** (AAVSO), 25 Birch Street, Cambridge, MA 02138 USA (*http://www.aavso.org/*) The AAVSO coordinates and collects the observations of approximately 600 observers from around the world through a variety of observing programs. Since the founding of the AAVSO in 1922, about 10 million observations of variable stars have been contributed to the AAVSO International Database by about 6000 observers. Variable star observations are collected, formatted, and submitted each month. There is a very specific format for reporting your observations and there are several ways to submit your reports to AAVSO Headquarters and several observation programs that encourage amateur participation are being conducted under the auspices of the AAVSO. The photoelectric monitoring of bright B[e] stars, the observation of small-amplitude red variables (SARVs), and PEP observation of RR Lyrae stars are just a few of the programs. Another great resource for amateurs since several professional astronomers are associated with AAVSO and are willing to assist those amateurs needing help.

Bundesdeutsche Arbeitsgemeinshaft fur Veranderlicht Sterne (BAV), Munsterdamm 90, 12169 Berlin, Germany (*http://thola.de/bav.html*) Founded in March 1950 by amateur astronomers in Berlin with the goal to collect, evaluate and record observations in Germany, BAV also produces ephemerides, charts, and publications. The BAV is organized into the following sections: 1) Evaluation and Publication, 2) Eclipsing binaries, 3) Short period stars, 4) Mira stars, 5) Pulsating stars, 6) Cataclysmic and eruptive stars, 7) Photoelectric and CCD-observation, and 8) Charts. Much of this Web site is in English but even if it wasn't, it's another great resource well worth your time.

Association Française des Observateurs d'Etoiles Variables (AFOEV), Observatoire Astronomique de Strasbourg, 11 rue de l'Université, 67000 Strasbourg (*http://cdsweb.u-strasbg.fr/afoev/*) The French Association of Variable Star Observers was founded in 1921 and the association headquarters is located at the Strasbourg Observatory. Currently, the association has about a hundred observers from 15 countries around the world. The observations made or received by the association has been published in their entirety in *Bulletin de l'AFOEV – 2nd series* (BAFOEV). They have been stored in the computer of the Centre de Données astronomiques at the Strasbourg Observatory. These observations now number more than 1,500,000 with the first recorded observation dating back to 1896. This database also includes observations made by other associations, including BAV (Germany), HAA (Hungary), NHK (Japan), NVVW (The Netherlands), Belgian, Norwegian, Swedish, Ukrainian and several astronomical groups in Spain. The observations are supplied free of charge. As with BAV, much of this Web site is in English and it will be worth your time to visit the AFOEV.

The **Variable Star Section of the Royal Astronomical Society of New Zealand** (RASNZ), PO Box 3181, Wellington, New Zealand (*http://www.rasnz.org.nz/*) The Royal Astronomical Society of New Zealand collates and coordinates southern hemisphere observations. By a long-standing reciprocal agreement, the RASNZ and BAAVSS exchange observational data on selected northern/southern hemisphere variables. The aim of the Society is the promotion and extension of knowledge regarding astronomy and other related branches of science. The RASNZ encourages an interest in astronomy, and as such is an association of observers and others providing mutual help and advancement of science. It was founded in 1920 as the New Zealand Astronomical Society and assumed its present title on receiving the Royal Charter in 1946. In 1967 it became a Member Body of the Royal Society of New Zealand. A great source of charts, data and other information for southern hemisphere stars. An excellent site and one you should visit.

BBSAG, M. Kohl, Im Brand 8, CH-8637 Laupen ZH (*mike.kohl@astroinfo.org*) The group of eclipsing binary observers of the Swiss Astronomical Society (BBSAG) acquires data of minima-times of eclipsing binary systems. Estimation of the magnitude itself is usually not required. The aim is to try to determine most exactly the time of maximum eclipse i.e. the time of minimum brightness. Long-term observations allow researchers to make statements on the evolution of such a system. Observations take several hours a night to cover an entire eclipse. A really excellent resource for eclipsing binaries.

The **International Supernova Network** (ISN) (*http://www.supernovae.net/isn.htm*) The ISN is strictly an Internet Web site. Its purpose is to share information among supernovae enthusiasts, both amateurs and professionals worldwide. The mailing list is used by members of the ISN to inform of newly discovered supernovae, to share observations, and to discuss topics concerning the search and observation of supernovae. The mailing list produces only a few messages per week but then supernovae are rare. You'll find everything that you need to hunt for supernovae at this Web site.

IAU: Central Bureau for Astronomical Telegrams (*http://cfa-www.harvard.edu/cfa/ps/cbat.html*) The *Central Bureau for Astronomical Telegrams* (CBAT) operates at the Smithsonian Astrophysical Observatory, under the auspices of Commission 6 of the *International Astronomical Union* (IAU) and is a nonprofit organization. The CBAT is responsible for the dissemination of information on transient astronomical events, via the IAU Circulars (IAUCs), a series of postcard-sized announcements issued at irregular intervals as necessary in both printed and electronic form. This Web site can be overwhelming when you first visit. Take your time, visit it several times. There is much of great value to be found here.

The Astronomer's Telegram (*http://atel.caltech.edu/*) The *Astronomer's Telegram* (*ATEL*) is for reporting and commenting upon new astronomical observations of transient sources. Content is limited to 4000 characters.

Chapter 15
Variable Star Observing and Amateur Astronomers

It is not the critic who counts; not the man who points out how the strong man stumbles, or where the doer of deeds could have done them better. The credit belongs to the man who is actually in the arena, whose face is marred by dust and sweat and blood; who strives valiantly; who errs, and comes short again and again, because there is no effort without error and shortcoming; but who does actually strive to do the deeds; who knows the great enthusiasms, the great devotions; who spends himself in a worthy cause; who at the best knows in the end the triumph of high achievement, and who at the worst, if he fails, as least fails while daring greatly, so that his place shall never be with those cold and timid souls who know neither victory nor defeat.

Theodore Roosevelt

Doubt has always existed among amateur astronomers regarding their contributions to the science of astronomy. Naturally, your equipment cannot compare with the professional's; you're unable to build a large telescope on some lofty mountain top and, if you're like the great majority of amateur astronomers, you lack the specialized training and experience of the professional astronomer; so the question of quality is posed. Occasionally, you may wonder, "Are my contributions even needed today?" when orbiting telescopes and 10-meter mirrors are being used by professional astronomers. The superiority of the professional and their equipment appears to leave not the slightest niche for you and me.

In reality, we have many advantages over our professional counterparts. We have access to a telescope whenever we wish, with due consideration for weather, family and social commitments. We have at our finger tips sophisticated equipment that until recently was only available to professional astronomers. As a result, we can produce quality data with very little notice and so we are able to make serious contributions to the science. Like the professional astronomer, we can possess indefatigable resolve, determination and love for the science of astronomy. However, the original question posed, the overall concern regarding valuable contributions, is a valid one.

In reality, the opportunities for us are growing, especially when the interest is variable stars. So as to assuage your concern, I am not attempting to compel you to make contributions to the science of astronomy. Your desire to do so will come naturally or it won't. My belief is that your desire to do so will evolve swiftly as your mind is drawn to the ever increasing wonder and awe that you will most assuredly experience and as the desire to share this new-found experience grows within you.

Without a doubt, the road on which you are about to travel will expose you to amazing sights but it can be rough and difficult; at times it may even appear to be impassable. However, variable-star observers are a tenacious, rugged lot. As a tribute to the amateur astronomer, in 1916, George Ellery Hale wrote:

Hampered, it may be, by lack of equipment, situated where the conditions for research are not of the best, and often compelled to devote <their> best hours to other pursuits, the amateur, rising above all discouragement has continued to pour a flood of new ideas and significant observations into the ever-widening sea of scientific knowledge.

A more stirring testimony to the devotion, determination and passion of the amateur astronomer will be difficult to find.

The history of amateur astronomy, especially variable-star observing, proclaims a strong tradition of contributions. The word "amateur" has its root in the Latin word "amator," or lover. The contemporary definition in vogue is "one who cultivates any art or pursuit for the love or enjoyment of it, instead of professionally or for gain." This definition, with emphasis on "pursuit for the love or enjoyment of it," certainly applies to variable-star observers. You're not going to make any money observing variable stars so if you do it, it must be for love.

The title of this chapter, in essence, refers to you and me. After reading the chapter title, you probably have formed within your mind's eye lists of variable stars, tables of data, detailed observing programs, and all sorts of complex and convoluted exercises developed to keep you occupied and interested. I leave all of that for you to develop. Really, much of the fun you're going to experience is in putting all of that together for yourself. You don't need me, or anyone else, forcing their own agenda down your throat. Refer back to Chapters 10 and 11 for guidance. I've provided you with just enough information so that you can develop your own observing programs or, if you desire, so that you may contact one of the variable-star organizations and request assistance. Chapter 12 provides sufficient guidance for you to begin observing variable stars. Nothing will be gained if I continue to tell you what to observe, when to observe it, or how to operate your equipment. All that you need now is to get out, under the stars, and begin. Start slow, eventually reach beyond your grasp, falter, strive, recover and move forward a little bit at a time. There really is no destination; it's all a journey.

So, in this regard I propose a handful of suggestions to help you during your investigations; some tools that will make you a better variable-star observer, a better amateur astronomer and a better scientist. Are you smiling? I do not use the word "scientist" loosely.

The definition to which I refer, "a scientific investigator," lays a heavy burden upon you. You must decide whether you will remain simply an observer, a humble spectator, one who watches from the side lines, in a sense just counting beans, occasionally excited by others to brief moments of enthusiasm but having no real impact on the overall adventure. Or perhaps, to decide to become a player, one of the team members on the field of play with a chance to make a real difference or at least to enjoy yourself by attempting difficult things. In any case, if you're serious, you're going to be investigating along scientific lines. You will, in fact, be a scientist. As such, while enjoying yourself, be serious about what you are doing. It's important.

Develop your target lists and tables of data, and prepare detailed observing programs. Read the current literature and make bold plans. And while doing so, especially during your serious study of variable stars, make an effort to consider the following suggestions:

Faithfully record important data and take meticulous notes. Both of these activities are acquired skills and both are different. Learn what is important and record it. You'll improve your skills by doing both and you'll get better as time passes. The difference between meticulous and verbose will become clear with time. Important information will soon become evident. And don't forget, not all of the important things will be seen through your eyepiece.

Follow the scientific method. If you're serious, then take yourself seriously so that others will also. Amateur sky divers are serious and take themselves seriously as do amateur cyclists, amateur scuba divers, amateur bird watchers, amateur gardeners and amateur cooks. As an amateur astronomer, you have an obligation to be serious about your work. The scientific method isn't perfect but it has worked well for those who have used it. Learn what it is, how it works and doesn't work, and apply it. You'll be surprised with your results. Believe that what you are doing is going to make a difference.

Do not jump to conclusions. Apply insight, logic and the scientific principle to describe your observations and subsequent analysis. Do not substitute the "law of parsimony" to make or defend a conclusion. Logical consistency and empirical evidence are absolute. Look for the simple answer. Simple revelations can be found within complicated and complex circumstances. Using a principle known as Occam's Razor, scientists realize that all things being equal, the simplest explanation is usually the correct one. The principle states that entities should not be multiplied unnecessarily. Albert Einstein plainly stated, "Everything should be made as simple as possible, but not simpler."

Submit proposals and results to peer review. You must learn to seek and accept "peer" review and criticism from other amateurs and from professionals of your observation programs, methods, and your results. This will be an important step in your continuous struggle to produce the highest quality results in all aspects of your activities. Peer review ensures that your science is sound, is consistent with prior knowledge found within the literature, and that your methodology and your results are repeatable by other competent individuals. Quite frankly, few amateur astronomers are willing to do this without arguing.

On the other hand, few professional astronomers are willing to take amateurs seriously enough to provide the needed attention and guidance with good explanations and gentle patience.

Fortunately, there are some within the professional astronomical community that selflessly help structure and guide amateur efforts: Dr. Arne Henden at the US Naval Observatory in Flagstaff, Arizona; Dr. Joe Patterson, Columbia University; Dr. John Percy, University of Toronto; Dr. Douglas Hall, Dyer Observatory and Dr. Taichi Kato, Kyoto University, come to mind. A handful of others exist too, but you must seek them out with due courtesy and respect.

Organizations and groups of variable-star observers will help you too. Organizations such as the *Variable Star Network* (VSNET); the *British Variable Star Section* (BAA VSS); the *American Association of Variable Star Observers* (AAVSO), the *International Amateur–Professional Photoelectric Photometry* (IAPPP); and the *Center for Backyard Astrophysics* (CBA) all exist to foster amateur–professional relationships. Look to them for assistance and guidance.

However, the greatest initiative must come from you. You have the most at stake because you are the one who may invest thousands of dollars on equipment, observing aids, and books with the expectation of serving science, and you will probably spend hundreds of hours observing each year in the hope that your results will make a real contribution to astronomy. For all of this to be productive, you, the amateur astronomer must strive for a new and stronger relationship with the professional community.

Reading a book seldom results in a competent understanding of any subject. I don't expect that this book will answer all of your questions or anticipate your every need and desire regarding variable-star observing. Now you need experience. Working through the challenges awaiting you will develop the experience that you need to feel comfortable when observing variable stars. In doing so, I don't believe that you're going to experience anything different than what all variable-star observers have experienced. In other words, don't feel intimidated. In the spirit of Tycho Brahe, William Herschel, F.W.A. Argelander, Ejnar Hertzsprung, Norman Pogson and many others, pick up your equipment and begin your quest! Focus your desire, your skill, and your determination. The absolute worst thing that can happen is that you will fail in some respect. In fact, I guarantee that you *will* fail in some regard. Overcoming failure is part of the overall challenge of variable-star observing and a failure is really a success, if viewed properly. When confused, ask questions and seek-out guidance; strive for what lies just beyond your grasp and attempt difficult things; take the time to understand what you are really doing and learn, don't just count beans.

Clear skies and good luck!

Database Management

Some types of databases will allow you to categorize, sort and retrieve the observations that you will collect over the years. Perhaps, in the beginning you may not feel a need for a computerized method, but as the years go by, you'll be surprised at the amount of information that you collect.

Within any database system, the field names are used to label each category of information, such as date, brightness estimate, type of observation, etc. Take a good look at your analysis needs and develop your field names with that requirement in mind. By doing so, you can move blocks of information from your database into your spreadsheet, for analysis, by simply using cut and paste. The reverse is also true; you will be able to move data from your spreadsheet into your database by using cut and paste.

Within the database, by carefully selecting your field names, you will be able to perform several valuable sort operations too. I recommend that you always prepare a master database, in which you enter your data. You may want a backup copy of this master file too. From the master database, prepare a working copy with which you can perform your sort operations. This way, if you make a mistake with the working copy, you can always go back to your master database and start again without losing any data.

A simple database, with field names, may look like this:

Date/time   Object   Type   Mag. Est.   Comp. star   Chart

Of course, you can make a database with many more fields but this simple example may serve as a good beginning. The "Date/time" should be your local date/time. When reporting variable star estimates, you can change to Julian date. The field name "Object" is used instead of simply "star" so that galaxy names can be entered when conducting supernovae searches. The "Type" allows you to indicate the type of variable star, such as Delta Scuti or CV, for example. The magnitude estimate should be based upon the "Comp. star" from the indicated "Chart."

Appendix B

Time Series Analysis Using TS11

Time series analysis allows you to search for periodic variability. Periodic variability means that the varying brightness of the star repeats with a good degree of precision. It's important to remember that not all variable stars are strictly periodic, even though they may vary in brightness. Cataclysmic variables vary in brightness and are usually said to have a "period" even though there can be great differences in time between their outbursts.

The program TS11, provided by the AAVSO, will allow you to make basic time series analysis of your variable-star data, and usually assist you in determining a period for some variable stars. In most cases, pulsating stars are the best candidates for getting the best results from this program

The most important thing to remember when using TS11 is that the data must be in columns, usually with the first column being the Julian date. The second column must be separated by one space; not a tab, or other such delimiter. The second column is usually the magnitude estimate. An example of a data file ready for TS11 is indicated next:

3456.123 6.5
3456.124 6.6
3456.125 6.7
3456.126 6.8

There is one space, and one space only, between the JD and the estimate. Because TS11 is a DOS program, the file name must be kept to eight characters or less. Be careful when you name your file, especially when you have many files containing data for the same star.

When you download your copy of TS11 from the AAVSO Web site, one of the files will be the instructions. Spend a couple of hours analyzing different data sets so that you

become familiar with the capabilities, and limitations, of this
program.

Appendix C

Analysis Using Spreadsheets

Graphs will allow you to better visualize your variable-star data. When using a computer to analyze columns of numbers, important information can certainly be discovered, but occasionally, using different methods to visualize your data will reveal subtle or cryptic details. In some cases, visualizing your data using several different methods will suggest alternate approaches for meaningful investigation. In any case, using graphs as a tool for the analysis of your data will expand your ability to extract useful data from your observations.

A good way to begin the analysis of variable-star data, using a spreadsheet, is to import the data directly from your database. By using a database and the cut and paste capabilities of most software, you need only enter your data once, usually into the database. Then you can move large blocks of data into your spreadsheets.

I recommend that you keep the field names the same within the database and spreadsheet. It will save time and reduce confusion. In some cases, you will want to expand your spreadsheet to include columns of calculated data that you do not wish to include within your database. A good way to keep your spreadsheet clean and uncluttered is to prepare one or more sections: perhaps one section to allow you to import your database information, then a section separated by a blank column in which you can perform your analysis calculations. Another section, slightly removed from the previous two, can be prepared to display your charts.

Remember, because your data is unevenly spaced, select your chart type carefully. Even a second or two, when analyzing fast variables such as RR Lyrae, Delta Scuti and eclipsing binaries, can produce a big effect on your analysis. Usually, an X-Y (scatter) chart is best used for the analysis of variable-star data because it displays uneven intervals, or clusters, of data. Usually, the time element is displayed along the X-axis and the brightness of the star (magnitude) is displayed along the Y-axis.

Experiment with displaying your data using slightly different methods and you may see how it produces different visual suggestions.

Index

Bold type indicates a major entry; *italic* type indicates an illustration.

Patrick Moore's Practical Astronomy Series

- of which Patrick Moore is the Series Editor - is for anyone who is seriously interested in the subject, but who may not be scientists, engineers or astronauts themselves!

All books in the series are written specifically for enthusiasts who have:

- gone beyond the first stages of learning about astronomy,

- who quite probably own, are thinking of buying, or have access to a moderately good optical telescope of some kind,

- and who want to read more.

All books are, of course, available from all good booksellers (who can order them even if they are not in stock), but if you have difficulties you can contact the publishers direct, by telephoning Freephone 00800 77746437 (in the UK and Europe),
+1/212/4 60 15 00 (in the USA),
or by emailing orders@springer.de

CALL US TODAY
FOR YOUR FREE
ASTRONOMY
BROCHURE

www.springer.de   www.springer-ny.com

3316